材料科学与工程专业系列教材

材料概论

Introduction to Materials

袁晓燕 李志宏 万怡灶 万小山 郭新权 摘编

天津大学出版社
TIANJIN UNIVERSITY PRESS

内 容 提 要

本书以各种材料的概述性基础知识及其重要品种的结构与应用为主要内容，涵盖金属材料、无机非金属材料、高分子材料及生物材料和复合材料。是继专业外语之后发展的一门材料类综合性双语教学课程教材。本书内容选自外文书籍、期刊和相关网站。全书共设5章，每章后附有关键词和简答题。

本书为高等理工科院校材料科学与工程专业本科公共基础理论课教材，同时也适用于本专业研究生的教学与科研，还可供从事材料方向研究和应用的科研技术人员参考。

图书在版编目(CIP)数据

材料概论/袁晓燕等摘编. —天津：天津大学出版社，2006.9 (2011.2 重印)

ISBN 978-7-5618-2342-2

Ⅰ.材... Ⅱ.袁... Ⅲ.材料科学－高等学校－教材－英、汉 Ⅳ.TB3

中国版本图书馆 CIP 数据核字(2006)第098436号

出版发行 天津大学出版社
出 版 人 杨欢
地　　址 天津市卫津路92号天津大学内(邮编：300072)
电　　话 发行部：022-27403647　邮购部：022-27402742
网　　址 www.tjup.com
短信网址 发送"天大"至916088
印　　刷 天津泰宇印务有限公司
经　　销 全国各地新华书店
开　　本 185mm×260mm
印　　张 10.5
字　　数 341千
版　　次 2006年9月第1版
印　　次 2011年2月第2次
印　　数 3 001－5 000
定　　价 16.00元

前　言

材料概论不仅包括高分子、无机非金属和金属三大材料，还涉及生物材料和复合材料的内容，是一门基于材料科学与工程一级学科的材料类概述性课程。

随着新的专业体制的调整，以往的三大材料专业——高分子材料专业、无机非金属材料专业和金属材料专业已合并为材料科学与工程专业。作为材料科学与工程专业下设的外文基础课程，专业外语已不能满足当前教学的需要。同时，随着材料科学的迅速发展，出现了许多新的材料品种，如纳米材料、生物降解材料等。为了适应新形势下材料专业的教学要求，培养学生灵活使用专业技术英语的能力，使之具备综合基本素质，我们结合以往的教学经验和科研背景，选取通俗易懂且知识新颖的内容，整理完成了这部《材料概论》双语课程教材。

本书是在天津大学"十五"规划重点课程建设项目的支持下完成的，是天津大学材料科学与工程专业平台课程《材料科学基础》、《材料物理性能》、《材料力学性能》、《材料分析方法》、《材料科学与工程实验》和《概料概论》系列教材的组成部分。不同于其他教材，本书以全英文的形式讲述高分子材料、陶瓷材料、金属材料以及复合材料的基础知识、各类材料的重要种类及其典型的应用实例。自2003年以来，本教材讲义已在本科生教学中使用过两次，并进行了多次修改。

全书共分5章，第1章介绍金属、陶瓷、高分子三大材料的结构特征，第2章至第5章分别介绍了金属、陶瓷、高分子和复合材料的重要类型及其应用。其中第1章由袁晓燕、李志宏、万小山、郭新权和万怡灶合作完成，第2章由万怡灶和万小山共同完成，第3章由李志宏完成，第4章由袁晓燕完成，第5章由万怡灶和郭新权共同完成。

在本书的整理过程中，李志崇教授审阅了全文，提出了有益的建议，同时也得到了原续波老师的指点，在此表示感谢！

由于本书的整理者水平有限，所选题材和内容可能存在片面性和不足，欢迎读者批评指正。

编者

2006年8月

Contents

1 Introduction

1.1 Historical perspective

Materials are probably more deep-seated in our culture than most of us realize. Transportation, housing, clothing, communication, recreation, and food production — virtually every segment of our everyday lives is influenced to one degree or another by materials. Historically, the development and advancement of societies have been intimately tied to the members' ability to produce and manipulate materials to fill their needs. In fact, early civilizations have been designated by the level of their materials development (i.e., Stone Age, Bronze Age).

The earliest humans had access to only a very limited number of materials, those that occur naturally: stone, wood, clay, skins, and so on. With time they discovered techniques for producing materials that had properties superior to those of the natural ones; these new materials included pottery and various metals. Furthermore, it was discovered that the properties of a material could be altered by heat treatments and by the addition of other substances. At this point, materials utilization was totally a selection process, that is, deciding from a given, rather limited set of materials the one that was best suited for an application by virtue of its characteristics. It was not until relatively recent times that scientists came to understand the relationships between the structural elements of materials and their properties. This knowledge, acquired in the past 60 years or so, has empowered them to fashion, to a large degree, the characteristics of materials. Thus, tens of thousands of different materials have evolved with rather specialized characteristics that meet the needs of our modern and complex society; these include metals, plastics, glasses, and fibers.

The development of many technologies that make our existence so comfortable has been intimately associated with the accessibility of suitable materials. An advancement in the understanding of a material type is often the forerunner to the stepwise progression of a technology. For example, automobiles would not have been possible without the availability of inexpensive steel or some other comparable substitute. In our contemporary era, sophisticated electronic devices rely on components that are made from what are called semiconducting materials.

1.2 Classification of materials

Solid materials have been conveniently grouped into three basic classifications: metals, ceramics, and polymers. This scheme is based primarily on chemical makeup and atomic structure, and most materials fall into one distinct grouping or another, although there are some intermediates. In addition, there are three other groups of important engineering materials — composites, semiconductors, and biomaterials. Composites consist of combinations of two or more different materials, where as semiconductors are utilized because of their unusual electrical characteristics; biomaterials are implanted into the human body. A brief explanation of the material types and representative characteristics is offered next.

Metallic materials are normally combinations of metallic elements. They have large numbers of nonlocalized electrons; that is, these electrons are not bound to particular atoms. Many properties of metals are directly attributable to these electrons. Metals are extremely good conductors of electricity and heat and are not transparent to visible light; a polished metal surface has a lustrous appearance. Furthermore, metals are quite strong, yet deformable, which accounts for their extensive use in structural applications.

Ceramics are compounds between metallic and nonmetallic elements: they are most frequently oxides, nitrides, and carbides. The wide range of materials that falls within this classification includes ceramics that are composed of clay minerals, cement, and glass. These materials are typically insulative to the passage of electricity and heat, and are more resistant to high temperatures and harsh environments than metals and polymers. With regard to mechanical behavior, ceramics are hard but very brittle.

Polymers include the familiar plastic and rubber materials and so on. Many of them are organic compounds that are chemically based on carbon, hydrogen, and other nonmetallic elements; furthermore, they have very large molecular structures. These materials typically have low densities and may be extremely flexible.

A number of composite materials have been engineered that consist of more than one material type. Fiberglass reinforced plastics is a familiar example, in which glass fibers are embedded within a polymeric material. A composite is designed to display a combination of the best characteristics of each of the component materials. Fiberglass acquires strength from the glass and flexibility from the polymer. Many of the recent material developments have involved composite materials.

Semiconductors have electrical properties that are intermediate between the electrical conductors and insulators. Furthermore, the electrical characteristics of these materials are extremely sensitive to the presence of minute concentrations of impurity atoms, which concentrations may be controlled over very small spatial regions. The semiconductors have made possible the advent of integrated circuitry that has totally revolutionized the electronics and computer industries (not to mention our lives) over the past two decades.

1.3 Structures of metals and ceramics

1.3.1 Structure of metals

One important feature which distinguishes a metal from most other materials is that in the solid state it is crystalline. This is another way of saying that its atoms are arranged in a regular three-dimensional pattern which has long-range order. Why should this be significant? Most of the properties which interest us in the use of metals in engineering are in some way related to the atomic arrangement. The properties obtained when metals are mixed together (i.e. in alloying) are dictated by the atomic structure. This also controls the way in which metals can be processed. Heat treatment and forming by plastic deformation are two examples of common metal-processing operations. How they can be used and the results achieved depend on the structure of the metal. If we are to understand how metals behave, we must be able to recognize the different arrangements of atoms which exist in commonly used metals.

Types of lattice structure

When discussing lattice structures, it is very convenient to think of atoms as solid spheres. We know this is not correct, of course; but, as we are interested only in the space occupied in the lattice by each atom, we do not need to consider its internal structure. This means that we can use spheres of, say, polystyrene or cotton wool to produce models which illustrate the various possible geometrical arrangements.

If we place a large number of spheres on a tray and line them up in rows, it can be seen that there are two basic arrangements. In the first (Figure 1.1(a)), the spheres are located at the corners of imaginary squares. Each sphere is in touch with four other spheres, and the tangents at points of contact are at right angles to each other. This is, therefore, a square arrangement. Alternatively, the rows can be displaced so that the spheres nestle between each other (Figure 1.1(b)). Each sphere is now in contact with six others which forms a hexagon around it. It will be noticed that there is less space between the spheres with this arrangement, and for this reason it is often referred to as close-packed.

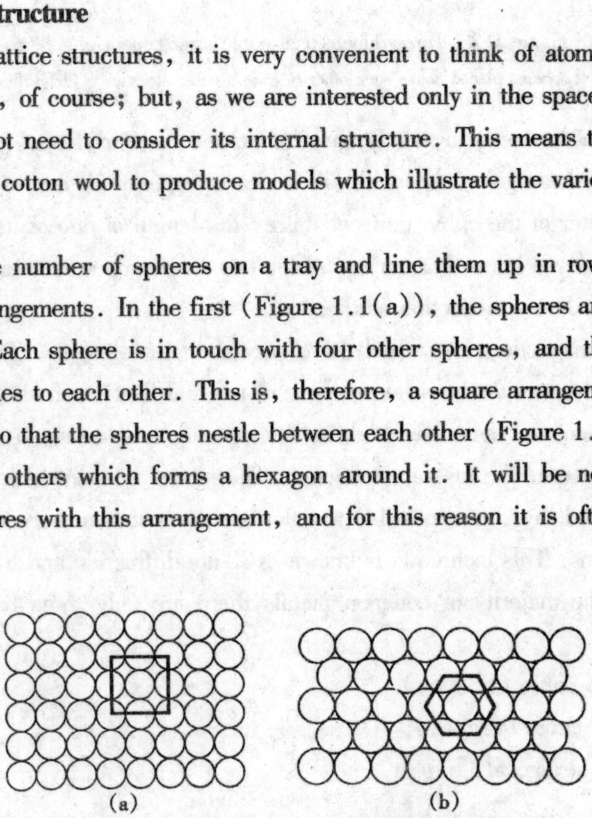

Figure 1.1 Two basic arrangements for spheres

(a) Square arrangement (b) Hexagonal arrangement

The next step in our investigation is to stack further layers of spheres on top of the first layer. With the square arrangement, we could position the spheres in the second layer immediately over those in the first, and so on with the third, fourth, and subsequent layers (Figure 1.2(a)). The result would be that the spheres would be sited at the intersections of a cubic space lattice (Figure 1.2(b)). It would also be possible to visualize the same arrangement being made up of a number of cubes stacked together. It follows that the unit cell, or basic "building block", for a lattice made up from a square arrangement is a simple cube. With a closed-packed arrangement, the unit cell is a hexagonal prism.

These two unit cells — cubic and hexagonal — feature in most of the lattice structures found in metals, although we must remember that the atoms, unlike our spheres, do not actually touch each

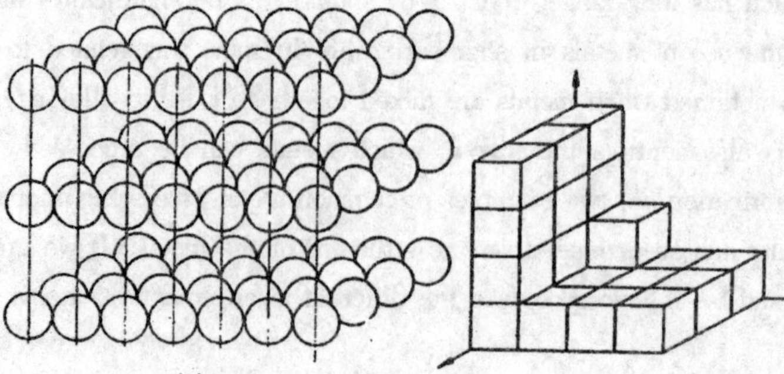

Figure 1.2 Three-dimensional cubic arrangement of spheres
(a) Layers of spheres placed above each other to form a cubic structure (b) Cubic space lattice

other, the dimensions of the cells in metals are, of course, vastly different from those of our models. The spheres we use for these give cubes with sides measuring from 20 mm to 50 mm. In the iron lattice, the lattice parameter of the cubic unit cell (i.e. the length of one of its sides) is about 2.86×10^{-10} m. How, then, do we know that our model lattices are reproduced at an atomic level? The main evidence has some from X-ray diffraction studies.

X-rays are electromagnetic waves which have very short wavelengths. They are able to penetrate metals and are frequently used to show the presence of internal flaws such as blow-holes in castings. If X-rays with wavelengths of the same magnitude as the interatomic spacings in a metal are directed at an angle to the planes of atoms, the rays are reflected. The planes thus act like a mirror reflecting light. By recording the reflected rays, it is possible to calculate the distances between the planes and to plot the positions of the atoms. This technique is known as X-ray diffraction and by its use it has been possible to show that in the majority of common metals there are only three basic lattice arrangements (Figure 1.3):

(a) body-centered cubic (b.c.c.)

(b) face-centered cubic (f.c.c.)

(c) close-packed hexagonal (c.p.h.)

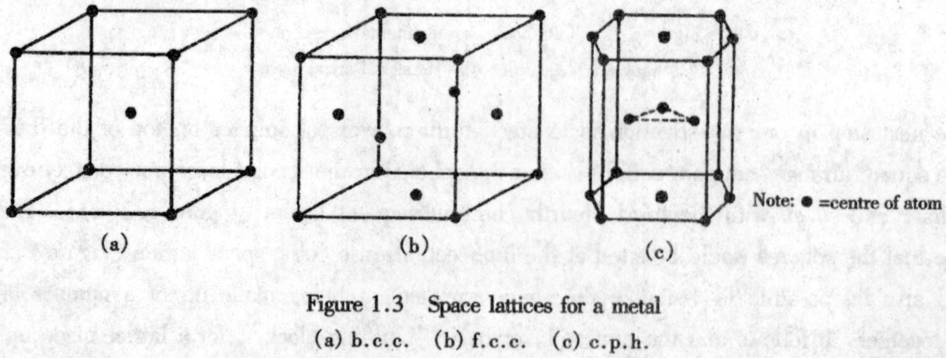

Figure 1.3 Space lattices for a metal
(a) b.c.c. (b) f.c.c. (c) c.p.h.

Body-centered cubic lattices have a cubic unit cell in which there is an atom at each corner and one in the center of the body of the cube. Although a b.c.c. lattice appears to be simple, it is not very often found in commercially available metals. Iron is the best-known metal which has a b.c.c.

lattice.

Face-centered cubic lattices also have an atom at the corner of each cube, but these are further apart than in the b.c.c. type. This leaves room for an additional atom at the center of each face of the cube. Many metals have f.c.c. lattices; examples are aluminum, copper, and nickel. In general, the most ductile metals have a f.c.c. structure.

Close-packed hexagonal lattices are significantly different from the cubic structures. The unit cell is not symmetrical. There are three layers in each cell. The top and bottom layers consist of six atoms in a hexagon with one atom at the center. The middle layer has three atoms in the form of a triangle. The vertical faces of the cell are rectangular. This is the least common of the lattice structures. Examples of metals with a c.p.h. lattice are zinc and magnesium — it is worth nothing that these are among the least ductile metals.

Imperfections in crystal lattices

At first sight, there is no reason to suppose that the lattice structure should not be uniform throughout its length, breadth, and depth. The positions which atoms occupy are determined by the forces acting on them. Since all the atoms in a metal are essentially uniform, it is reasonable to expect that interatomic forces are the same at all points in the lattice. In practice, however, imperfections do occur. This may be because there was some interruption in the growth of the crystal from molten metal. Imperfections can also be created by the inclusion of an atom of another metal, whether added intentionally or present as an impurity. They are worthy of note because they can have a significant effect on the behavior of a metal.

An important type of imperfection is a dislocation (Figure 1.4). As its name implies, this is a break in the continuity of the lattice. Very often a plane of atoms will come to a stop. At this point the two neighboring planes move closer until they are at the correct interatomic spacing. It looks almost as if one plane has been squeezed out. The change from three to two planes causes a disruption of the lattice and leaves a small gap; this is called an edge dislocation. Dislocations are an essential element in the forming of metals since they allow plastic deformation to take place at achievable stress levels.

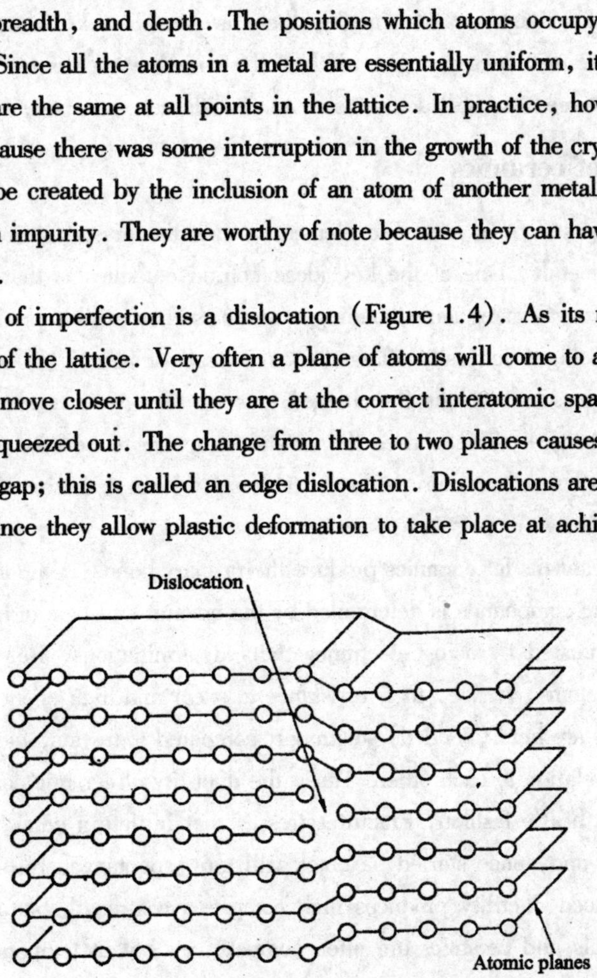

Figure 1.4 Dislocation in a crystal lattice

Another form of imperfection is a vacancy. This is literally a hole in the lattice which is there because an atom did not take up the prescribed site. Vacancies play an important role in the diffusion or movement of atoms through the lattice.

Grain structure

It is possible to produce rods of metal in which the orientation or direction of the lattice planes remains the same throughout the body of the specimen. Each plane goes from one side of the rod to the other without change in direction. Similarly along the length, continuity is maintained from one end to the other. These are identified as single crystals. They are very useful in studying the basic properties of a metal, but they are very difficult to produce and are not found in general manufacture. The metals which we use are polycrystalline, i.e. they are made up of a number of small crystals or grains bonded together by interatomic forces. Within each grain, the lattice structure is uniform.

Grains do not have the regular surface appearance we normally associate with crystals. Their shape is determined by interaction with neighboring grains and can be quite irregular. If we took a slice through a piece of metal — say copper — the boundaries of the grains would appear as a network of lines. The grain boundary represents a change in the direction of the lattice planes. As the planes of one grain meet those of a neighbor at an angle, there is considerable mismatch with the result that along the boundary there may be gaps. The width of the boundary area is usually equivalent to several atomic diameters, but forces can still act across it to provide cohesion between the grains.

1.3.2 Structure of ceramics

Ceramics are composed of at least two elements, thus the crystal structures are generally more complex than those of metals. One of the key ideas behind ceramics is that ceramics exhibit atomic bonding through either covalent or ionic bonding. Ceramics that exhibit predominantly ionic bonding can be thought of as being electrically charged ions instead of atoms. In ionic bonding of ceramic structure, the metallic ions, or cations (+ charge) give up their valence electrons to the nonmetallic ions, or anions (– charge). This causes the crystal structure to become electrically neutral. One common example of this type of structure is that of table salt (NaCl), in this case there are equal number of cations and anions.

The bonding mechanisms for ceramics produce the primary bonds: ionic and covalent. The crystal structure of bulk ceramic compounds is determined by the amount and type of bonds. The percentage of ionic bonds can be estimated by using electronegativity determinations. Being compounds, ceramics have different types of atoms. Hence, their resistance to shear and high-energy slip is extremely high.

Because the atoms are held (bonded) so strongly compared to metals, there are fewer ways for atoms to move or slip in relation to each other. Thus, the ductility of ceramic compounds is very low and these materials act in a brittle fashion. Fracture stresses that initiate a crack build up before there is any plastic deformation and, once started, a crack will grow spontaneously. The combination of high shear stresses and reduced ductility produces high compressive strength but low tensile strength. At room temperature, metals and ceramics are often competitive, but at temperatures above 1,500 °F, metals weaken while ceramics retain much of their strength. Ceramics are noted for their heat resistance. The maximum service temperature for alumina is 3,450 °F and for silicon carbide, it is 3,000 °F. Heat-resistant nickel alloys are considered unserviceable above 1,500 °F.

The porosity of ceramics plays a major role in the strength. Ceramics are generally very porous, enhancing its ability to insulate but limiting its tensile strength.

1.4 Polymer structures

Naturally occurring polymers, those derived from plants and animals, have been used for many centuries; these materials include wood, rubber, cotton, wool, leather, and silk. Other natural polymers such as proteins, enzymes, starches, and cellulose are important in biological and physiological processes in plants and animal. Modern scientific research tools have made possible the determination of the molecular structures of this group of materials, and the development of numerous polymers, which are synthesized from small organic molecules. Many of our useful plastics, rubbers, and fiber materials are synthetic polymers. In fact, since the conclusion of World War II, the field of materials has been virtually revolutionized by the advent of synthetic polymers. The synthetics can be produced inexpensively, and their properties may be managed to the degree that many are superior to their natural counterparts. In some applications metal and wood parts have been replaced by plastics, which have satisfactory properties and may be produced at a lower cost.

1.4.1 Macromolecules

The molecules in polymers are gigantic in comparison to the hydrocarbon molecules; because of their size they are often referred to as macromolecules. Within each molecule, the atoms are bound together by covalent interatomic bonds. For most polymers, these molecules are in the form of long and flexible chains, the backbone of which is a string of carbon atoms; many times each carbon atom singly bonds to two adjacent carbon atoms on either side, represented schematically in two dimensions as follows. Each of the two remaining valence electrons for every carbon atom may be involved in side-bonding with atoms or radicals that are positioned adjacent to the chain. Of course, both chain and side double bonds are also possible.

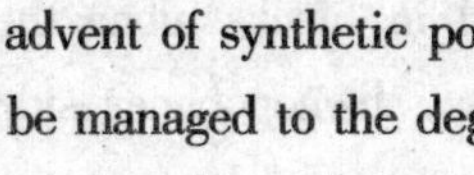

These long molecules are composed of structural entities called mer units, which are successively repeated along the chain. "Mer" originates from the Greek word meros, which means part; the term polymer was coined to mean many mers. Generally, we use the term monomer, which refers to a stable molecule from which a polymer is synthesized. The polymer molecule contains a structural identity, repeating itself several times. These repeating entities are called repeat units or monomeric units of the polymer molecule.

When all the repeating units along a chain are of the same type, the resulting polymer is called a homopolymer. There is no restriction in polymer synthesis that prevents the formation of compounds other than homopolymers; and, in fact, chains may be composed of two or more different monomeric units, in what are termed copolymers.

Polymer chemists and scientists are continually searching for new materials that can be easily and economically synthesized and fabricated, with improved properties or better property combinations than those offered by the homopolymers. One group of these materials is the copolymer. Consider a copolymer that is composed of two monomeric units as represented by ○ and ● in Figure 1.5. Depending on the polymerization process and the relative fractions of these monomer types, different sequencing arrangements along the polymer chains are possible. For one, as depicted in Figure 1.5(a), the two different units are randomly dispersed along the chain in what is termed a random copolymer. For an alternating copolymer, as the name suggests, the two monomeric units alternate chain positions, as illustrated in Figure 1.5(b). A block copolymer is one in which identical monomers are clustered in blocks along the chain (Figure 1.5(c)). And, finally, homopolymer side branches of one type may be grafted to homopolymer main chains that are composed of a different monomer; such a material is termed a graft copolymer (Figure 1.5(d)).

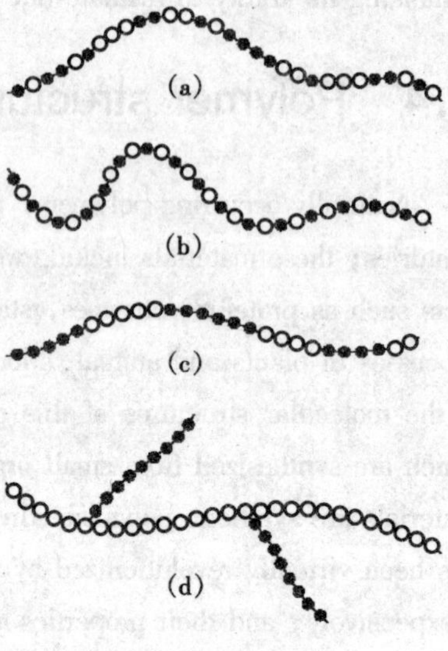

Figure 1.5 Schematic representations of (a) random, (b) alternating, (c) block, and (d) graft copolymers. The two different monomer types are designated by open and solid circles

1.4.2 Molecular weight

Extremely large molecular weights* are to be found in polymers with very long chains. During the polymerization process in which these large macromolecules are synthesized from smaller molecules, not all polymer chains will grow to the same length. This results in a distribution of chain lengths or molecular weights.

Ordinarily, an average molecular weight is specified, which may be determined by the measurement of various physical properties such as viscosity and osmotic pressure. There are several ways of defining average molecular weight. The number-average molecular weight $\overline{M}_n$ is obtained by dividing the chains into a series of size ranges and then determining the number fraction of chains within each size range. The weight-average molecular weight $\overline{M}_w$ is based on the weight fraction of molecules within the various size ranges. A typical molecular weight distribution along with these molecular weight averages are shown in Figure 1.6.

An alternate way of expressing average chain size of a polymer is as the degree of polymerization, which represents the average number of monomeric units in a chain.

Various polymer characteristics are affected by the magnitude of the molecular weight. One of these is the melting or softening temperature; melting temperature is raised with increasing molecular weight (for $\overline{M}$ up to about 100,000 g/mol). At room temperature, polymers with very short chains (having molecular weights on the order of 100 g/mol) exist as liquids or gases. Those with molecular

* "Molecular mass", "molar mass", and "relative molecular mass" are sometimes used and are really more appropriate terms than "molecular weight" in the context of the present discussion — in actual fact, we are dealing with masses and not weights. However, molecular weight is most commonly found in the polymer literature, and thus will be used throughout this book.

weights of approximately 1,000 g/mol are waxy solids (such as paraffin wax) and soft resins. Solid polymers (sometimes termed high polymers), which are of prime interest here, commonly have molecular weights ranging between 10,000 and several million g/mol.

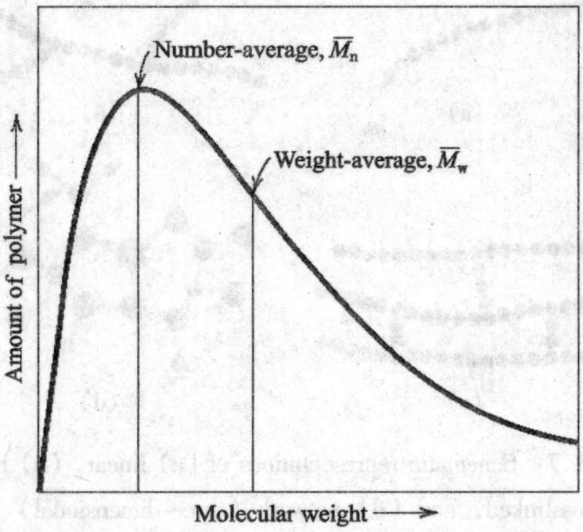

Figure 1.6 Distribution of molecular weight for a typical polymer

1.4.3 Molecular structures

The physical characteristics of a polymer depend not only on its molecular weight and shape, but also on differences in the structure of the molecular chains. Modern polymer synthesis techniques permit considerable control over various structural possibilities. This section discusses several molecular structures including linear, branched, crosslinked, and network.

Linear polymers are those in which the monomeric units are joined together end to end in single chains. These long chains are flexible and may be thought of as a mass of spaghetti, as represented schematically in Figure 1.7(a), where each circle represents a monomeric unit. For linear polymers, there may be extensive van der Waals and hydrogen bonding between the chains. Some of the common polymers that form with linear structures are polyethylene, poly(vinyl chloride), polystyrene, poly(methyl methacrylate), nylon, and the fluorocarbons.

Polymers may be synthesized in which side-branch chains are connected to the main ones, as indicated schematically in Figure 1.7(b); these are fittingly called branched polymers. The branches, considered to be part of the main-chain molecule, result from side reactions that occur during the synthesis of the polymer. The chain packing efficiency is reduced with the formation of side branches, which results in a lowering of the polymer density. Those polymers that form linear structures may also be branched.

In crosslinked polymers, adjacent linear chains are joined one to another at various positions by covalent bonds, as represented in Figure 1.7(c). The process of crosslinking is achieved either during synthesis or by a nonreversible chemical reaction that is usually carried out at an elevated temperature. Often, this crosslinking is accomplished by additive atoms or molecules that are covalently bonded to the chains. Many of the rubber elastic materials are crosslinked; in rubbers, this is called vulcanization.

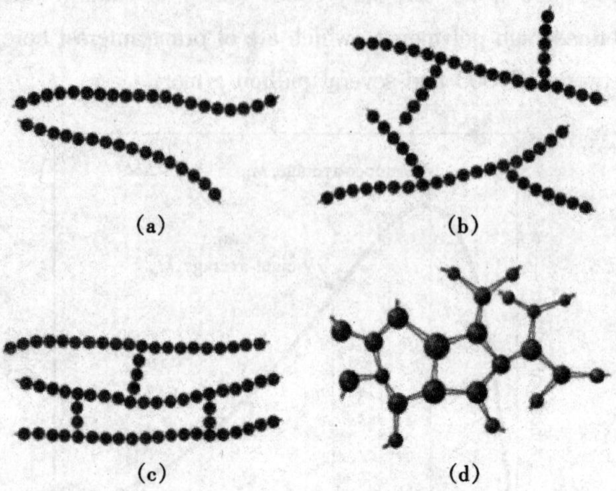

Figure 1.7 Schematic representations of (a) linear, (b) branched, (c) crosslinked, and (d) network (three-dimensional) molecular structures. Circles designate individual monomeric units

Trifunctional monomeric units, having three active covalent bonds, form three-dimensional networks (Figure 1.7(d)) and are termed network polymers. Actually, a polymer that is highly crosslinked may be classified as a network polymer. These materials have distinctive mechanical and thermal properties; the epoxies and phenol-formaldehyde belong to this group.

It should be pointed out that polymers are not usually of only one distinctive structural type. For example, a predominantly linear polymer might have some limited branching and crosslinking.

Keywords

alternating copolymer	body-centered cubic	brittle fracture
block copolymer	branched polymer	close-packed hexagonal
crystal structure	covalent bond	ceramics
copolymer	crosslinked polymer	dislocation
degree of polymerization	face-centered cubic	grain boundary
graft copolymer	homopolymer	ionic bond
lattice structure	linear polymer	macromolecule
monomer	molecular weight	molecular structure
network polymer	random copolymer	

Problems

1. How many basic lattice arrangements are there in the majority of common metals, and what are the differences among them?

2. What are main bonding mechanisms between atoms in ceramics?

3. Describe a typical polymer molecule in terms of its chain structure, and how the molecule may

be generated by repeating monomeric units.

4. What are number-average and weight-average molecular weights, and degree of polymerization for a specified polymer?

5. Name and briefly describe the four general types of polymer molecular structures.

Selected from

W. D. Callister. Fundamentals of Materials Science and Engineering. 5th Edition. John Wiley & Sons, 2001

2 Metal Materials

2.1 Structure and properties of steel

Of all the factors which contributed to the success of the industrial revolution in the last century, one of the most important must surely have been the availability of steel in commercial quantities. There is no record in history of the actual discovery of steel, but swords and knives made from metal have been valued for centuries because of their sharp cutting edges which did not blunt rapidly in use. These same attributes were essential for the tools which were needed in the industrial revolution to machine drive shafts, gears, and other engineering components. The growth of industry and the transport system also created a need for a more adaptable material than cast iron. Designers looked for a strong ductile metal which they could use in structures such as ships, trains, bridges, and boilers, and found that low-carbon steel (originally known as mild steel) satisfied their requirements. In addition, methods developed by Bessemer and Siemens enabled this steel to be manufactured in large quantities at relatively low prices.

Steel is so commonplace in our everyday lives that we often overlook the fact that there are many types, each designed by the metallurgist to have some particular properties required by the engineer or designer. The requirements for a steel vary considerably but more often than not involve consideration of tensile properties and/or impact strength. Some examples of typical applications with their associated requirements are listed in Table 2.1.

Table 2.1 Typical carbon steels

Application	Requirements	Composition (%)				
		C	Mn	Si	P	S
Car body panels	Sheet which can be pressed accurately into shape, especially around sharp corners. A suitable steel must have good ductility with a reasonable low yield stress.	0.08	0.3	–	0.05	0.05
Ships' hulls	Plates must have good strength, to withstand the stresses experienced in service in heavy seas. There is also a need for good impact strength at sub-zero temperatures, to avoid catastrophic brittle fracture. The steel used must also be weldable, since all modern ships' hulls are fabricated by welding.	0.18	0.8	0.1	0.05	0.05

Table 2.1 Typical carbon steels **(Continued)**

Application	Requirements	Composition (%)				
		C	Mn	Si	P	S
Axle shafts	The bar stock used for this application must have good strength in bending and torsion. It is also often desirable to harden the surface layers to improve resistance to wear.	0.4	0.8	0.1	0.05	0.05
Helical springs	The steel must first be capable of being rolled into rods. The next requirement is good ductility, to enable the coils to be formed. Finally the steel must respond to heat treatment to give suitable mechanical properties.	1.0	0.6	0.3	0.05	0.05

If we look at the compositions of the steels which meet these requirements, we observe that they all contain carbon, manganese, silicon, phosphorus, and sulphur. The last two elements are impurities introduced by the raw materials used to make the steel. The steelmaker tries to keep these as low as possible. Normally the sulphur and phosphorus contents do not exceed 0.05% and we do not need to take them into consideration. The small amounts of silicon are residues from the steelmaking process and do not directly influence the properties of the steels listed in Table 2.1. We are left, therefore, with carbon and manganese. Both of these alloying elements have a significant influence on the mechanical properties of the steel, but the more important is the carbon. The amounts of carbon in steel are small when compared with the alloy contents of the non-ferrous metals which we will discuss in another chapter. Nevertheless, even small variations in the carbon content have a profound effect not only on the properties of the steel but also on the way in which these can be altered by heat treatment.

2.1.1 Iron-carbon alloys

The first important fact to note about steels is that they are part of a range of iron-carbon alloys (Figure 2.1). As we can see, iron and carbon combine together to form a compound, iron carbide. In this, one carbon atom is bonded to three iron atoms; hence it is given the simple formula Fe_3C, although its crystal structure is complex. When we are talking about the structure of iron-carbon alloys, the carbide is usually called cementite.

The amount of cementite present in an iron-carbon alloy depends on the carbon content. If there is little or no carbon present, the microstructure consists of uniform grains of ferrite with no cementite and the metal is soft and ductile. Ferrite is the name given to pure body-centered-cubic iron with a very small amount of carbon in solution. At the other end of the scale, with a carbon level of 6.67%, the structure consists entirely of cementite: the material is very hard with virtually no ductility and no commercial use. Between these two extremes, the ductility depends on the proportion of cementite in the microstructure and hence on the carbon content (Table 2.2).

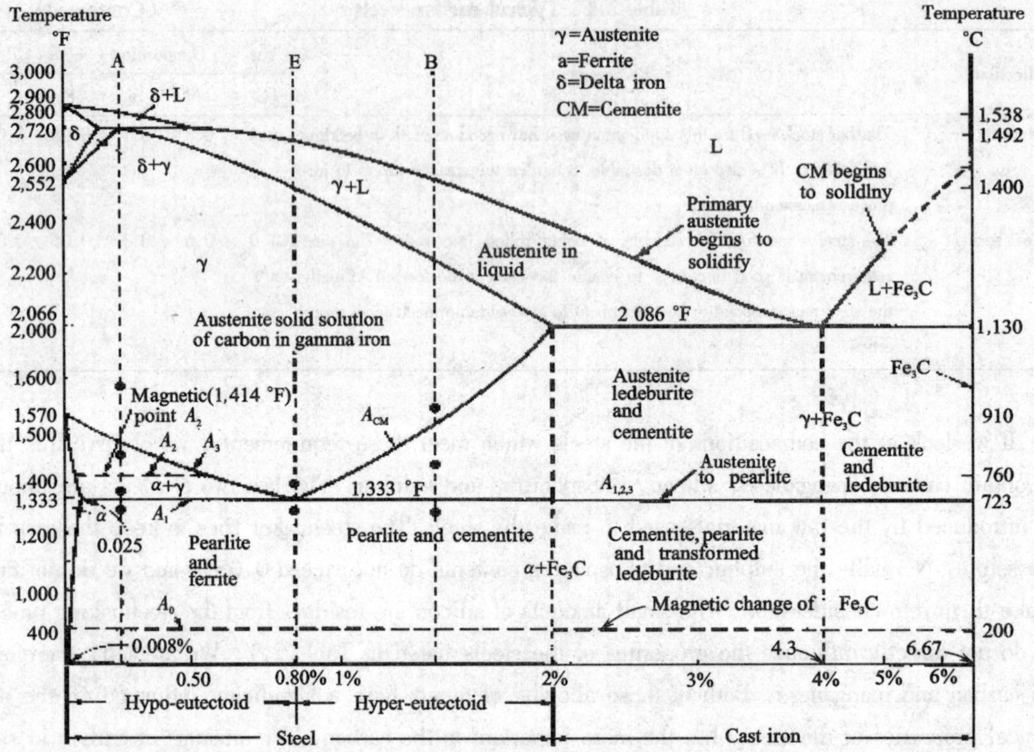

Figure 2.1 Iron-carbon phase diagram

Table 2.2 Ductility of plain carbon steels

Carbon content (%)	Elongation in tensile test (%)
Nil	42
0.2	37
0.4	31
0.6	22
0.8	17
1.2	3

Note: carbon-manganese steels would give lower elongations than those quoted above.

Based on carbon content, iron-carbon alloys can be divided into four groups which offer different combinations of properties and characteristics:

(a) 0.1% to 0.8% carbon — steels for general engineering;

(b) 0.9% to 1.2% carbon — special steels, e.g. for wear resistance;

(c) 1.3% to 2.2% carbon — not normally used;

(d) 2.0% to 4.2% carbon — cast iron.

In this chapter, we will be concerned mainly with those steels which are in common use in engineering manufacture (i.e. group (a)).

2.1.2 Strength versus carbon content in steels

The role of carbon can be seen most readily in the range of strengths which are available from var-

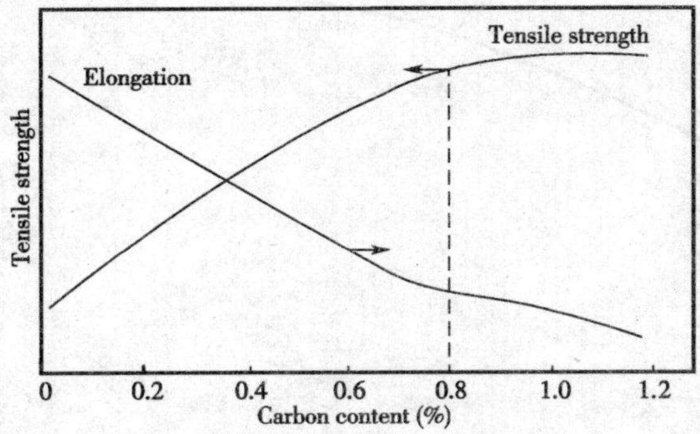

Figure 2.2 Relationship between strength and carbon content

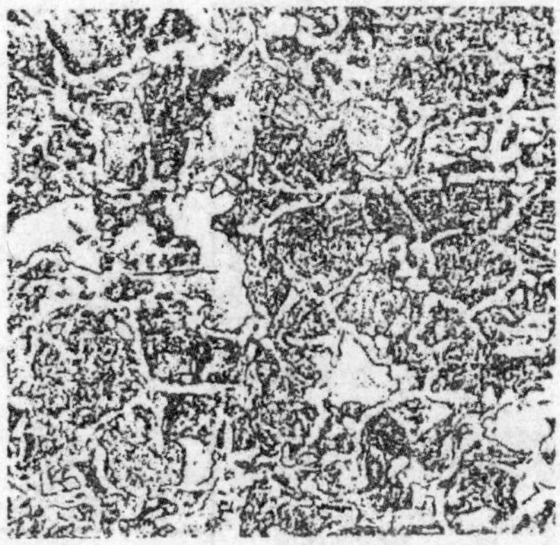

Figure 2.3 Microstructure of a 0.4% carbon steel after slow cooling (magnification × 150)

ious steels (Figure 2.2). As the carbon content is increased from zero to 0.8%, the tensile strength goes from 250 N/mm^2 to 850 N/mm^2. These changes in strength must be related to the microstructure. As we have already seen, at very low carbon contents the metal consists entirely of grains of ferrite. At 0.4% carbon, when the strength is about 540 N/mm^2, just over half the area viewed through a microscope is occupied by a constituent known as pearlite (Figure 2.3). In a steel containing 0.8% carbon, the metal is completely pearlitic and the maximum strength is reached (Figure 2.4).

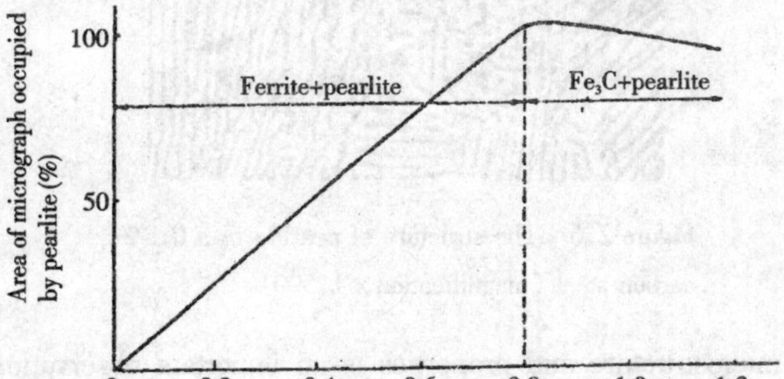

Figure 2.4 Relationship between pearlite % and carbon content

It follows that there is some relationship between the presence of pearlite and the increase in the strength of the steel. By plotting tensile strength as a function of the percentage of the area occupied by pearlite (Figure 2.5), we can see that the relationship is similar to the one shown in Figure 2.2 between strength and carbon content. The link is to be found in the structure of pearlite. At low magnifications, etched pearlite often appears to be a uniform material with the colors of mother-of-pearl — this is the origin of its name. Examination at higher magnifications (Figure 2.6) shows that it has a laminated structure. It is a composite consisting of alternate layers of ferrite and cementite. Since the amount of cementite in a steel is determined by the carbon content, it follows that the primary effect of carbon in relation to strength is seen in the amount of pearlite it produces. In other words, increases in carbon content act indirectly through the changes which they cause in the microstructure.

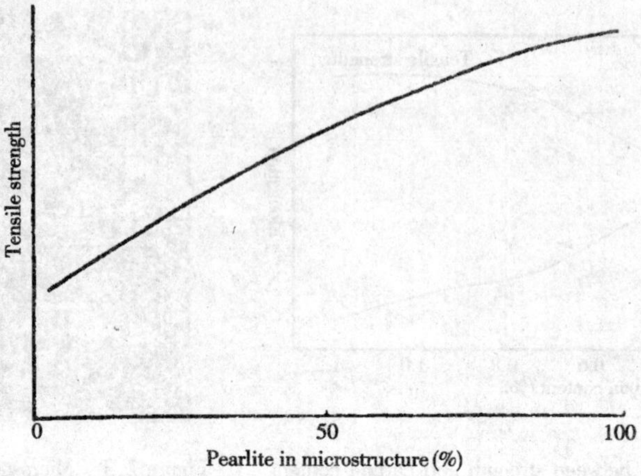

Figure 2.5 Relationship between pearlite % and strength

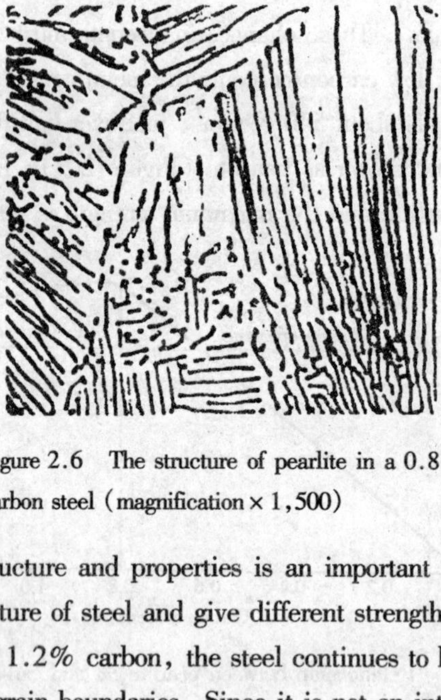

Figure 2.6 The structure of pearlite in a 0.8% carbon steel (magnification × 1,500)

This link between microstructure and properties is an important observation since heat treatment can be used to change the structure of steel and give different strengths with the same carbon content.

Within the range 0.9% to 1.2% carbon, the steel continues to have a pearlitic structure but cementite forms a network at the grain boundaries. Since it is not an integral component of the pearlite, it is referred to as free cementite. This network of a hard brittle compound at the grain boundaries is associated with an increase in the hardness, but it causes a marked deterioration of the ductility and this is undesirable for most commercial purposes.

In general, the strength does not increase with carbon content between 0.9% and 1.2%. At best, it remains that of the fully pearlitic structure, but it may even fall to a value below that of the 0.8% carbon steel.

2.1.3 Formation of pearlite

As pearlite is so obviously a critical constituent in the structure of a steel, it is important to understand how it is formed and controlled. The formation of pearlite, and indeed the whole metallurgy of steels, relies on the allotropy of iron. A metal is allotropic if it can exist in two or more crystalline forms.

Allotropic forms of iron

Iron has two allotropic forms which have different atomic structures. The first of these, α iron, has a body-centered-cubic lattice and exists at room temperature. When raised to elevated temperatures, the atoms re-arrange themselves and adopt a face-centered-cubic arrangement, γ iron. On cooling, the structure reverts to body-centered-cubic, the change taking place at 910 ℃. These two forms of iron have different physical properties, e.g. coefficient of expansion. Energy is expended in the change from γ to α iron, and the cooling curve shows a thermal arrest at the temperature at which the event occurs. The term transformation or reaction is often used when referring to this change from one allotropic form to the other.

Transformation of austenite

During the cooling of a piece of pure iron, the transformation occurs at a unique temperature (910 ℃). Above this point the metal is all in the γ form (i.e. face-centered-cubic lattice), while below it the entire crystal structure is body-centered cubic (i.e. the α form).

If carbon is added to the iron to produce a steel, the change from γ to α takes place over a range of temperatures. The limits of this range are known as the critical points and, by convention, the higher temperature — i.e. the one at which the transformation starts during cooling — is designated A_3 while the lower point is A_1. The A_1 temperature remains constant at 723 ℃ irrespective of the carbon content. On the other hand, the A_3 temperature is progressively reduced as the carbon content is increased, until it reaches 723 ℃ at 0.8% carbon. Above this level, the A_3 temperature rises again.

Plotting the A_3 and A_1 temperatures for a number of steels with carbon contents up to 1.2% produces an equilibrium diagram which is part of the larger iron-carbon diagram (Figure 2.7). It shows the transformation under equilibrium cooling conditions of austenite, which is the name given to γ iron

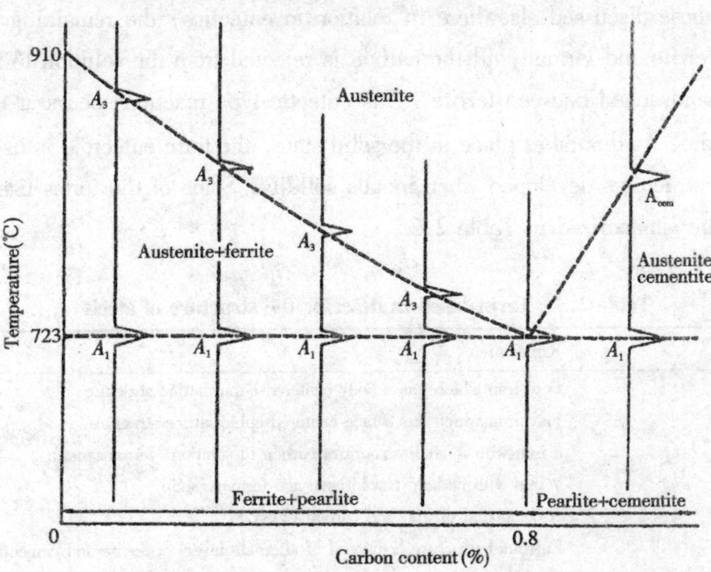

Figure 2.7 Critical points for carbon steels

with carbon in solution. Although the diagram shows a solid-to-solid transformation, we can apply the rules of interpretation which we used for equilibrium diagrams concerned with solidification.

The first point we need to note is the difference in carbon solubility between austenite and ferrite. The face-centered-cubic lattice can hold as much as 1.7% carbon in solution at 1,130℃. Thus, with the range of steels we have been discussing, the structure is entirely austenitic at this temperature and all the carbon is in solution. The carbon is dissolved interstitially: in other words, a carbon atom does not replace an iron atom to form part of the cube but nestles between two iron atoms (Figure 2.8). When the lattice is rearranged to give a body-centered-cubic pattern, there is inadequate space between the iron atoms and the carbon atom is ejected from its site. This is the reaction which takes place at the A_3 point.

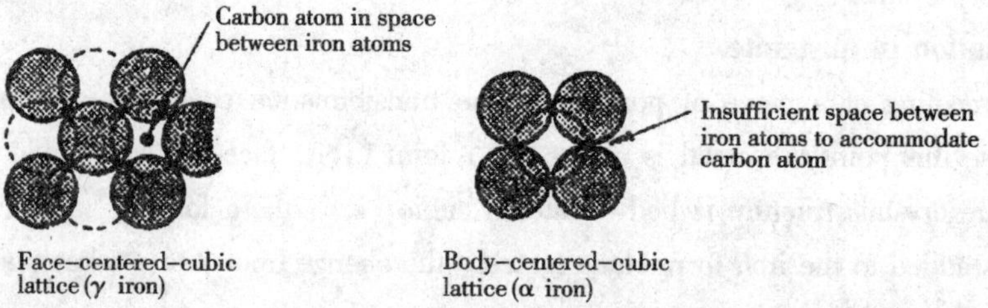

Figure 2.8 Interstitial solution of carbon

Looking at the event diagrammatically, we can visualize the b.c.c. ferrite grain being nucleated at the edge of an austenite grain (Figure 2.9) with the result that a boundary is set up between the two constituents. As the temperature is lowered, the boundary advances further and further into the austenite grain. When the carbon is rejected from the ferrite lattice, it diffuses into the body of the austenite grain and is retained in solution. This means that the untransformed austenite becomes progressively richer in carbon until the amount of this element in solution reaches 0.8%. At this point the events are very similar to those discussed elsewhere in relation to eutectics: the remaining austenite is completely changed to ferrite and virtually all the carbon is rejected from the solution to form layers of cementite which are sandwiched between ferrite. This eutectic-type reaction occurs at the A_1 point, and the product is pearlite. As this takes place in the solid state, the term eutectoid is used, to distinguish it from the eutectic structures developed when metals solidify. Some of the terms used to describe the structure of steels are summarized in Table 2.3.

Table 2.3 Terms used to describe the structure of steels

Term	Explanation
α iron (alpha iron)	Pure iron which has a body-centered-cubic lattice structure.
γ iron (gamma iron)	Pure iron which has a face-centered-cubic lattice structure.
Ferrite	α iron with a small amount of carbon in solution; ferromagnetic.
Austenite	γ iron with carbon in solution; non-ferromagnetic.
Cementite	A compound of iron and carbon (Fe_3C).
Pearlite	Eutectoid structure composed of alternate layers of ferrite and cementite.
Hypo-eutectoid steels	Steels which have a carbon content less than that of the eutectoid (0.8%).
Hyper-eutectoid steels	Steels with a carbon content above the eutectoid composition.
Martensite	A hardened structure produced by quenching a steel from above the A_3 point.

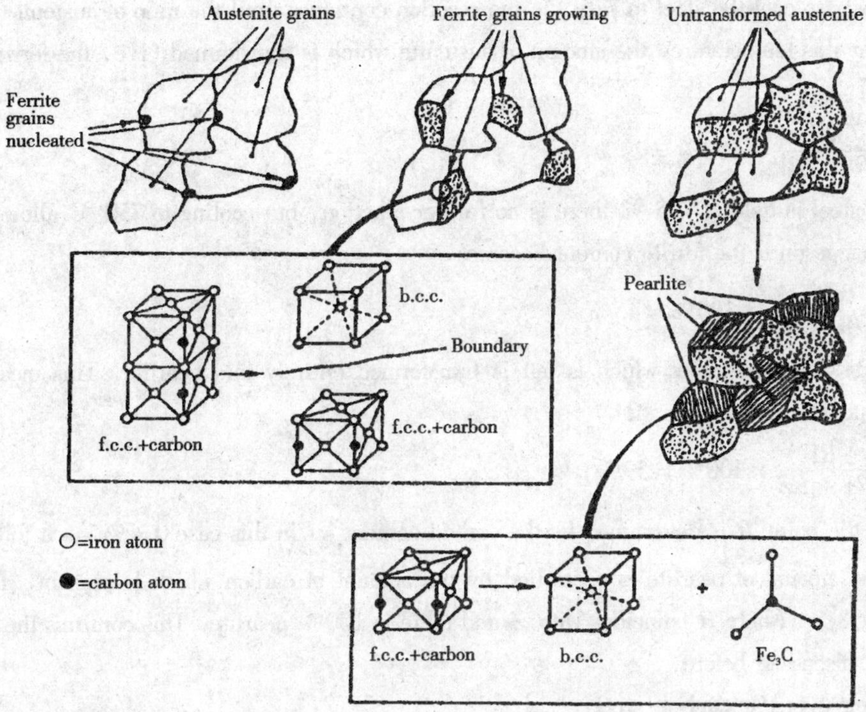

Figure 2.9 Transformation of austenite

Pearlite content

The proportion of austenite which transforms into pearlite can be determined from the equilibrium diagram (Figure 2.10).

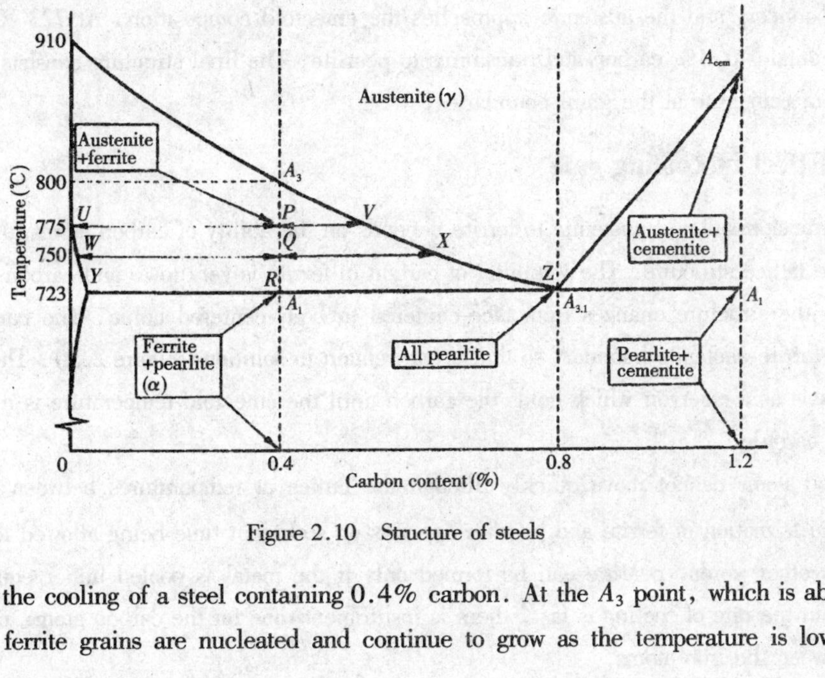

Figure 2.10 Structure of steels

Consider the cooling of a steel containing 0.4% carbon. At the A_3 point, which is about 800 ℃ for this steel, ferrite grains are nucleated and continue to grow as the temperature is lowered. The amount of austenite which is changed to ferrite is determined by the temperature.

Suppose we cool the steel to 775 ℃, the reaction continues until the ratio of austenite to ferrite is *PU*/*PV*. At this temperature, the amount of austenite which is transformed (i.e. the ferrite content) is

$$\frac{PV}{PU + PV} \times 100\% = 29\%$$

If the steel is held at 775 ℃ there is no further reaction, but cooling to 750 ℃ allows more austenite to change until the ferrite content becomes

$$\frac{QX}{QW + QX} \times 100\% = 45\%$$

At 723 ℃ the austenite which is left is transformed entirely into pearlite. This means that the pearlite content of the steel is

$$\frac{RY}{RY + RZ} \times 100\% = 59\%$$

Since the point *R* is determined by the carbon content — in this case 0.4% — it follows that in any steel the amount of pearlite is controlled by the amount of carbon which is present. The limiting case is at 0.8%, where *R* coincides with *Z* and there is 100% pearlite. This confirms the microscopic evidence discussed before.

Cooling a 1.2% carbon steel

With this steel, the austenite at 1,000 ℃ contains more carbon in solution than the eutectoid composition. At the onset of the transformation, at approximately 875 ℃, carbon is again rejected from solution but in this case forms free cementite at the grain boundaries. The austenite grain is thus depleted of carbon. For clarity, this temperature is referred to as the A_{cem} point, to distinguish it from the austenite-to-ferrite transformation point (A_3). As the temperature falls, more cementite is deposited at the boundary and the austenite approaches the eutectoid composition. At 723 ℃ the remaining austenite contains 0.8% carbon and transforms to pearlite. The final structure consists of pearlite with a network of cementite at the grain boundaries.

2.1.4 Effect of cooling rate

The transformation of austenite to ferrite depends on the ability of carbon atoms to move or diffuse through the lattice structure. The solubility of carbon in ferrite is very low, and carbon is rejected from solution as the structure changes from face-centered to body-centered cubic. The carbon must move across the ferrite-austenite boundary so that it can remain in solution (Figure 2.9). The untransformed austenite acts as a reservoir which holds the carbon until the eutectoid temperature is reached, when it forms iron carbide.

Carbon atoms cannot move quickly through the lattice at temperatures between the A_1 and A_3 points. The formation of ferrite and pearlite depends on sufficient time being allowed for the carbon to diffuse. In other words, pearlite can be formed only if the metal is cooled under equilibrium conditions. When the rate of cooling is fast, there is insufficient time for the carbon atoms to leave their positions between the iron atoms.

The distance between the corner atoms in the face-centered lattice is 23% greater than in the body-centered-cubic cell. If a carbon atom is trapped between two iron atoms, they are kept apart and are not able to take up the equilibrium positions for a body-centered cube. The result is that the lattice is distorted. Dislocations cannot move freely in a distorted lattice, and the steel becomes harder and loses its ductility. This is what happens when a chisel is hardened by heating it to just above the A_3 temperature and then plunging it into water or oil (i.e. quenching). The quenching gives a fast cooling rate and reduces the temperature to about 300 ℃ before the carbon has had time to diffuse. The austenite-to-ferrite transformation is suppressed and the material is permanently hard, since carbon diffusion rates are negligible below 300 ℃ and the carbon atoms stay fixed between the iron atoms. The lattice structure remains distorted and can be rectified only if the temperature is raised again to allow the carbon to diffuse. Under the microscope the structure looks acicular or needle-like and is known as martensite (Figure 2.11).

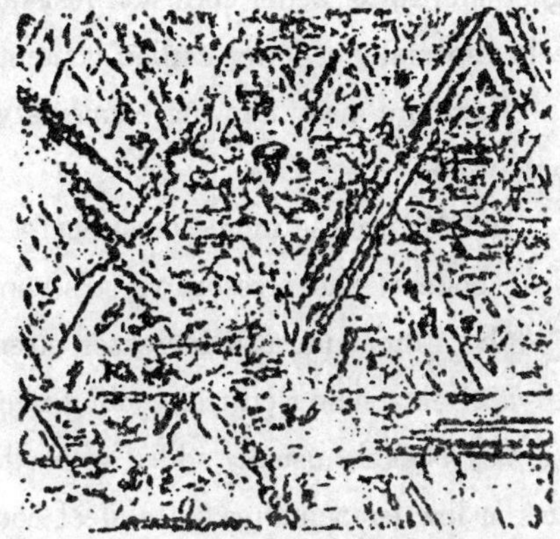

Figure 2.11 Martensitic structure in a 0.4% carbon steel (magnification × 500)

The increase in hardness achieved by quenching depends on the carbon content. At low carbon levels there is very little change. The effect begins to be more noticeable at about 0.4% carbon and reaches a maximum at the eutectoid composition. The increases at 0.4% and 0.8% carbon are 255 HV and 530 HV, respectively.

Since the formation of martensite depends on the cooling rate, we need to know how fast the steel must be cooled to produce hardening. This is the critical cooling rate and for carbon steel it is somewhere between 400 ℃/s and 500 ℃/s. Carbon content affects the critical cooling rate. As the carbon content increases, the critical rate becomes slower (Figure 2.12). This means that it is easier to harden an 0.8% carbon steel than one containing 0.4%.

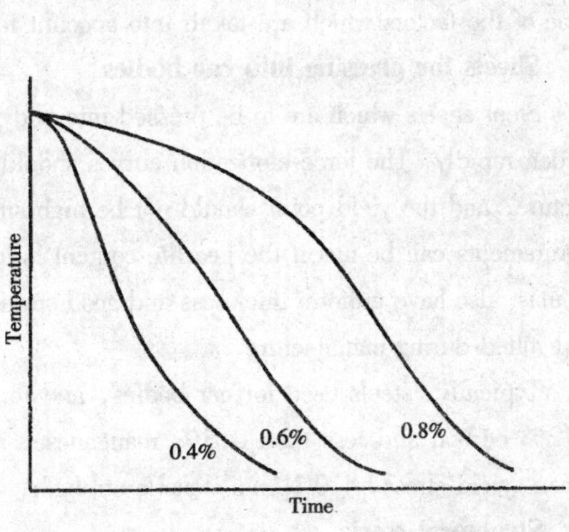

Figure 2.12 Critical cooling rates for different carbon contents

In normal heat-treatment practice, quenching into water represents about the fastest cooling rate we can achieve. This corresponds to the critical rate for an 0.35% carbon steel. Hence steels with this carbon content or less are not suitable for applications where the component is to be quench-hardened.

2.1.5 Alloying elements

Frequently alloying elements are added to carbon steels to achieve specific properties, such as higher strength or better corrosion resistance. Sometimes the aim may be to make it easier to harden the steel, especially at lower carbon contents.

There are many alloying elements which can be added, but the following are the four most commonly used:

(a) Manganese increases strength and hardness. Most steels used for structural work, bridges, ships, boilers, and pressure vessels contain between 0.8% and 1.6% manganese.

(b) Chromium also increases strength and hardness, and makes it easier to harden the steel, i.e. it gives a slower critical cooling rate. Additions in excess of 13% confer good corrosion resistance.

(c) Nickel, usually added with other elements, increases strength and improves the impact properties at low temperatures — a steel containing 9% nickel is used for storage tanks containing liquid methane at − 162 ℃.

(d) Molybdenum in amounts between 0.5% and 1.0% improves creep strength. Normally it is added together with 1.5% to 5% chromium.

Although silicon is occasionally used as an alloying element, it is mainly added to the molten steel ingot to reduce the oxygen content. In most steels a residual amount of silicon is present but it does not contribute to the properties.

2.1.6 Selecting carbon steels

Steels offer a wide variety of properties which must be matched to service or manufacturing requirements if the best grade is to be selected. The following examples have been chosen to illustrate some of the factors which are taken into account in typical applications.

Sheets for pressing into car bodies

Steel sheets which are to be pressed into body panels must have good ductility and must not work-harden rapidly. The force-elongation curves should show an appreciable amount of plastic strain before fracture, and the yield point should not be high since this could call for a more powerful press. These requirements can be met if the pearlite content is low and the steel is almost entirely ferritic. The sheets must also have uniform thickness and good surface finish, which means that the steel will have been cold rolled during manufacture.

Typically, steels used for car bodies, and for similar applications, contain between 0.07% and 0.12% carbon and less than 0.5% manganese. Corresponding mechanical properties would be 210 N/mm^2 yield stress, 410 N/mm^2 tensile strength, 44% elongation, and 61% reduction of area.

Structural work

Structures such as bridges, building frames, and storage tanks are generally designed on the basis of the yield stress. Quite often the maximum allowable design stress is two thirds of the yield stress, and a steel must be chosen which meets this requirement. However, the choice of a steel is frequently limited by the need to be able to form the plates or sections by bending, and to weld them together. This sets an upper limit to the carbon content, and manganese is added to bring the strength up to the desired level. British Standard BS 4360: 1979 covers a range of steels which are suitable for most structural applications. These are divided into four grades giving yield stresses ranging from 230 to 450 N/mm^2 (Table 2.4). These steels usually contain 0.14% to 0.2% carbon and 0.8% to 1.2% manganese. The relative amounts of these two elements are adjusted by the steelmaker to give the required mechanical properties.

Table 2.4 Structural steels (BS 4360:1979)

Grade	Yield strength (N/mm^2)	Tensile strength (N/mm^2)	Elongation (%)
40	230 — 260	400 — 480	22
43	240 — 280	430 — 510	20
50	340 — 355	490 — 620	18
55	415 — 450	550 — 700	17

Note: the grades are further subdivided according to Charpy impact values, for example:

Grade	Temperature of specimen during test (℃)	Minimum Charpy test value (J)
40A	Not specified	Not specified
40B	Room temperature	27
40C	0	27
40D	−10	41
	−20	27
40E	−20	61
	−30	47
	−50	27

Frequently, in structural applications there is also a requirement for good impact properties, especially at temperatures around freezing point and below. The steels in BS 4360 are further classified according to Charpy value.

Casting

Metals for casting must have acceptable fluidity. The fluidity in steels is not good but can be improved by the addition of about 0.6% silicon. If it is to be used for casting, a steel must have high resistance to cracking, which means that the carbon content should not be high. A number of carbon steels which are suitable for casting are listed in BS 3100:1979(see Table 2.5). These have carbon contents of less than 0.45%, and increased strength is achieved by manganese additions.

Table 2.5 Carbon-steel castings for general purposes (from BS 3100:1979)

Grade	Composition (%) (max.)			Tensile strength	Elongation (%)
	Carbon	Manganese	Silicon	(N/mm^2) (minimum)	(min.)
A1	0.25	0.9	0.6	430	22
A2	0.35	1.0	0.6	490	18
A3	0.45	1.0	0.6	540	14
A4	0.25	1.6	0.6	540	16
A5	0.33	1.6	0.6	620	13

Note: the castings can be heat-treated to achieve the specified properties.

Machine components

A large variety of machine parts such as gear wheels, shafts, connecting rods, agricultural tools, and turbine discs are made by forging and machining. Steels for these applications must have not only adequate strength at room temperatures but also high ductility at forging temperatures (750 ℃ to 1,200

℃). A further requirement may be for heat treatment to improve wear resistance or toughness. Generally steels for this type of application contain 0.35% to 0.6% carbon and are typified by BS 970: 1972 (see Table 2.6).

Table 2.6 Mechanical properties of medium-carbon steels (from BS 970: part 1: 1972)

	Steel designation		
	'40' carbon (080 M40)	'50' carbon (080 M50)	'55' carbon (070 M55)
Carbon (min.)(%)	0.32	0.45	0.50
(max.)(%)	0.40	0.55	0.60
Manganese (min.)(%)	0.60	0.60	0.50
(max.)(%)	1.00	1.00	0.90
Yield stress (N/mm^2) (min.)	275	310	355
Yield strength (N/mm^2) (min.)	540	620	695
Elongation(%)	16	14	12

Note: properties quoted are for the normalized condition.

Some steels tend to tear during machining, reducing cutting speeds and giving a poor surface finish. Better results can be obtained by using a steel which contains 0.1% to 0.2% sulphur or lead. Free-cutting steels, as they are called, are available in a range of carbon contents as bar or rod. Their main disadvantage is that they crack when arc welded, and they are weak at the solidification temperature. Studding which is to be welded into place and shafts which have welded flanges must be made from steels with a lower sulphur content, even though this means some sacrifice of machinability. The alternative is to use a process such as friction welding or flash welding which does not involve melting.

2.2 Heat treatment of steels

The dependence of the structure of steel and cast iron on the cooling rate opens up the possibility of using heat treatment to produce a range of properties.

2.2.1 Response of steels to heat treatment

Three heat treatments are commonly applied to hypo-eutectoid carbon steels: annealing, normalizing, and hardening. In each of these, the steel is heated to 50 ℃ above the A_3 point (Figure 2.13). The structure is austenitic at this temperature, and all the carbon is taken into solution. The properties of the steel depend on how fast it is cooled from the austenitic condition to room temperature. Essentially, the difference between the three heat treatments listed is the rate of cooling from the A_3 point.

Full annealing is an approach to equilibrium conditions. The steel is cooled in the furnace so that the cooling rate can be controlled by gradually reducing the heat supplied, thus allowing the γ – to – α transformation to go to completion at each temperature. The structure consists of large grains of ferrite with coarse pearlite in which there are thick plates of ferrite and carbide. This gives the softest possible structure and is an ideal preparation for mechanical working. Low yield stress and tensile strength are associated with annealing.

Normalizing is used as a finishing treatment for carbon steels as it gives higher strength than annealing without serious loss of ductility. The steel is removed from the furnace and allowed to cool in air, so that the cooling rate is faster than in annealing. The final grain size is smaller than with annealing because a larger number of grains are nucleated and there is less time for them to grow.

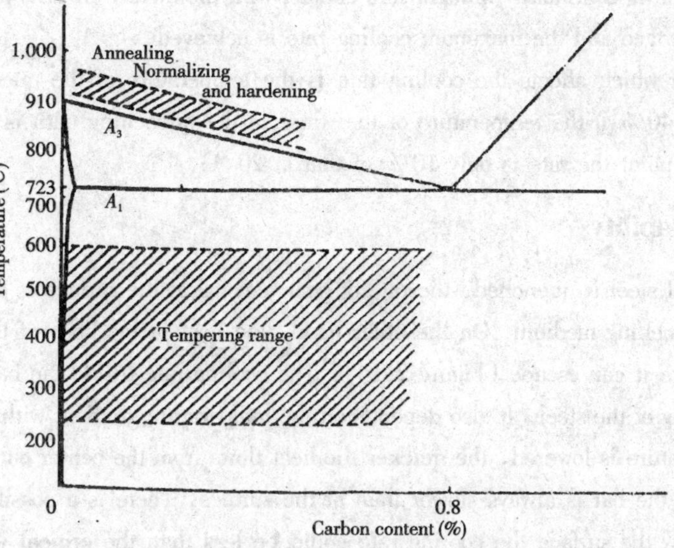

Figure 2.13 Heat-treatment ranges for carbon steels

At low carbon contents, say 0.2%, the effect of this small grain size is more noticeable in impact strength than in tensile strength. The latter increases by only about 5% compared with annealed material, but impact strength (Charpy value) can be improved by as much as 20% by normalizing. Steel plates are often delivered in the normalized condition when high impact-test values are specified, especially at low temperatures, e.g. BS 4360 grades 40D and 40E (see Table 2.4). At 0.4% carbon, the effect on strength also becomes marked, and improvements of 15% in the yield stress are obtainable.

Hardening relies on exceeding the cooling rate needed to produce a martensite structure. This involves removing the steel from the furnace and plunging it immediately into a quenching medium.

There is little advantage to be gained from cooling faster than the critical rate — on the contrary, distortion and cracking occur with very fast cooling. The quenching medium is therefore chosen to cool at a rate which just exceeds the critical value. Three commonly used quenching media are listed in Table 2.7; salt solution (brine) gives the fastest cooling rate. Brine would be used for a 0.35% carbon steel which is difficult to harden, while oil would be suitable for a eutectoid steel.

Table 2.7 Quenching media

Quenching medium	Relative cooling rate
5% salt solution (in water)	1.0
Water at 20 ℃	0.85
Quenching oil	0.42
Mineral oil	0.30

The effectiveness of the medium depends to some extent on the techniques used for quenching. As soon as the steel component is plunged into the quenching bath, the liquid in contact with the surface of the metal is vaporized. A layer of vapor is established over the surface which effectively insulates the steel and slows down the cooling rate. The component must be agitated to break up this film and to ensure that fresh liquid is continually brought into contact with the metal. In this way, heat is conducted rapidly from the surface and the maximum cooling rate is achieved.

Another factor which affects the cooling rate is the temperature of the quenching medium. The rate is reduced by 40% if the temperature of the water in the quenching bath is raised from 20 ℃ to 40 ℃. At boiling point the rate is only 10% of that at 20 ℃.

2.2.2 Hardenability

When a bar of steel is quenched, the rate of heat removal at the surface is governed by the effectiveness of the quenching medium. On the other hand, the heat in the body of the bar must first flow to the surface where it can escape (Figure 2.14). The cooling rate inside the bar is controlled by the thermal conductivity of the steel. It also depends on the temperature gradient within the bar. The faster the surface temperature is lowered, the quicker the heat flows from the center outwards. However, the cooling rate within the bar is always slower than at the surface. There is a possibility, therefore, that at some point below the surface the cooling rate could be less than the critical value, with the result that the metal would not harden.

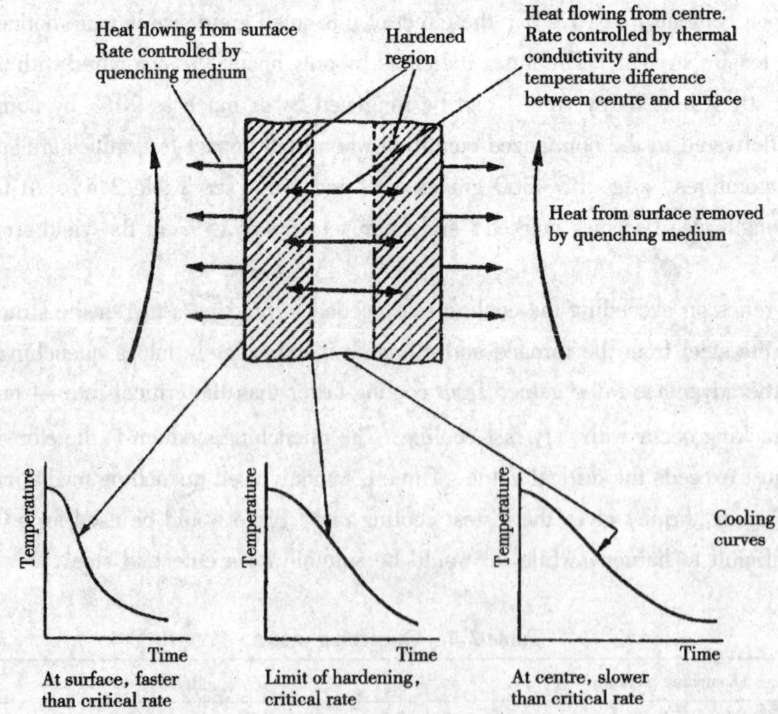

Figure 2.14 Depth of hardening in steel

The hardenability of steel is a measure of the depth of hardening when it has been quenched. If a steel has good hardenability, even thick material has uniform hardness through the complete cross-section. By contrast, a bar made from a steel with low or poor hardenability has a soft center (or core) after quench hardening.

The hardenability of a steel depends on three prime factors:

(a) The composition of the steel — steels which have a low critical cooling rate have better hardenability; hence steels with a high carbon content harden to a greater depth than low-carbon steels. Alloying elements such as chromium improve the hardenability of a steel.

(b) The quenching medium.

(c) The section dimensions (size of component).

The hardenability of a steel is indicated in Standards by stating the ruling section. This is the maximum thickness which can be used and still achieve the stated properties through the complete section. In other words, it is the maximum depth to which the steel hardens by a specified amount on quenching from above the A_3 temperature.

Size is also important in relation to distortion. Variations in the rate of cooling within the component can lead to different amounts of contraction at various points through the thickness. One part of the component can contract more than another, and the shape or dimensions can be changed. To some extent this can be minimized by the technique used to lower the part into the quenching bath. Thus with thin sections the long axis should be vertical. In this way, the heat flows quickly through the thin section with the result that the temperature distribution is relatively uniform.

2.2.3 Surface hardening

In some applications there is a need for the surface to be hard, to resist wear, while the body of the component remains ductile. There are two ways of hardening the surface of a steel component. In the first method, a thin surface layer is raised to above the A_3 point and is then quenched. Either oxygen-fuel-gas burners (flame hardening) or induction coils can be used to heat the surface. The method is only suitable for steels with more than 0.4% carbon, which means that the strength of the unaffected metal in the body of the component will be relatively high.

As an alternative, the composition of the surface layers of a ductile low-carbon steel can be altered to make them responsive to quench hardening. One variant of this technique (case hardening) involves enriching the surface with carbon. Steels used for case hardening contain about 0.2% carbon, and the aim is to increase this to between 0.8% to 0.9% to a depth of 0.5 mm to 1 mm. To achieve this, the component must be heated in a carbon-rich atmosphere. It can be packed into a box containing charcoal, or immersed in a bath of fused sodium-cyanide/soda-ash mixture, or kept in a methane atmosphere during heating. With an 0.2% carbon steel, the component must be heated to 900 ℃ to 950 ℃ and held at this temperature for about three hours while carbon diffuses into the surface layers. The component is then quenched to form martensite in this carburized metal. As the core is still low in carbon, it does not harden. Sometimes a second heat treatment may be necessary to adjust the properties of the core. For example, the grain size of the low-carbon-steel core may increase during the prolonged heating at 900 ℃, with a consequent loss of mechanical properties. A second heating and quenching can reduce the grain size in the core.

2.2.4 Tempering

Steel sections which have been hardened by quenching are brittle and contain stresses which result to some extent from one part of the component cooling faster than another. The transformation from autstenite to martensite also creates internal stress. These stresses can be reduced to a very low level by tempering. This consists of heating the hardened component to between 200 ℃ and 600 ℃ and holding it at temperature for a predetermined period of time. Tempering also alters the mechanical properties.

At the lower end of the temperature range there is very little effect on the structure but some of the stresses are relieved. Above 230 ℃, carbon atoms which were trapped in the lattice during quenching move from these non-equilibrium positions and form small particles of cementite. The strain in the lattice is reduced by the removal of the carbon atoms, and the hardness of the martensite is lowered. This is accompanied by an increase in the toughness and impact strength of the steel. The extent of the softening is a function of the temperature used for the tempering operation — the amount of softening is greater at higher temperatures. As the temperature is raised from 230 ℃, progressively more carbon is released form the lattice, the cementite particles become larger, and both the hardness and the strength fall.

The steel must be held at temperature long enough to allow all the related changes to take place. Tempering temperatures must not exceed the A_1 point, as the steel would become partially austenitic and the benefits of the hardening treatment would be lost.

In heat-treatment practice, the tempering temperature is related to the function of the component. Tools are treated within the range 230 ℃ to 300 ℃. From 230 ℃ to 260 ℃ the aim is to give a lasting cutting edge with good abrasion resistance. Tempering between 270 ℃ and 300 ℃ confers better shock resistance and springiness while retaining a good cutting edge. Traditionally the temperature is judged by the color of the oxide film which forms on the surface of the steel (Table 2.8).

Although temper colors have been widely used for many years, more accurate methods of temperature measurement are to be preferred.

If greater ductility and toughness are required, for example in shafts and high-strength bolts, the steel is tempered at 300℃ to 600℃.

Table 2.8 Tempering temperatures for tools

Temperature (℃)	Color of oxide on surface	Type of tool
230	Pale straw	Planning tools
240	Dark straw	Milling cutters
250	Brown	Taps and dies
260	Purplish-brown	Punches; drill bits
270	Purple	Press tools
280	Dark purple	Cold chisels
300	Blue	Wood saws; springs

2.2.5 Choice of heat treatment

The choice of the most suitable heat-treatment sequence is dictated by what is to happen to the steel. If it is to be subjected to further processing, either recrystallization or full annealing would be appropriate. On the other hand, heat treatment during the finishing stages is often specified to ensure that properties match specification. Quite often this is covered by a Standard. The information from BS 3100:1976 in Table 2.9 illustrates the way that the manufacturer is guided in the choice of heat treatment while still being allowed some latitude where properties are easily achieved (grades A4 and A5).

Table 2.9 Properties and heat treatment of steel castings (from BS 3100:1976)

1.5% manganese steel castings for general purposes

Chemical composition The steel shall contain:

Element	A4 % min.	A4 % max.	A5 and A6 % min.	A5 and A6 % max.
Carbon	0.18	0.25	0.25	0.33
Silicon	–	0.60	–	0.60
Manganese	1.20	1.60	1.20	1.60
Phosphorus	–	0.050	–	0.050
Sulphur	–	0.050	–	0.050

Heat treatment All castings shall be supplied in the heat-treated condition, the heat treatment being in accordance with the requirements specified in (a) to (c) below.

(a) Steel A4 The final heat treatment shall consist of normalizing, normalizing and tempering, or hardening and tempering at suitable temperatures to give the mechanical properties specified.

(b) Steel A5 The final heat treatment shall consist of normalizing, normalizing and tempering, or hardening and tempering at suitable temperatures to give the mechanical properties specified for this steel, which is suitable for castings having section thicknesses of up to 100 mm.

(c) Steel A6 The final heat treatment shall consist of oil or water hardening and tempering at suitable temperatures to give the mechanical properties specified for this steel which, in this condition, is suitable only for castings having section thicknesses of up to 63 mm. If greater section thicknesses are involved and similar mechanical properties are required, it is necessary that a steel of higher hardenability, e.g. steel BT1, be specified.

Mechanical properties The mechanical properties to be obtained from test-pieces selected, prepared and tested in accordance with the requirements of section 1 of this standard shall be as follows.

Mechanical properties	A4	A5*	A6+
Tensile strength			
N/mm^2 (min.)	540	620	690
N/mm^2 (max.)	690	770	850
Lower yield stress or 0.2% proof stress, N/mm^2 (min.)	320	370	495
Elongation, % (min.)	16	13	13
Charpy V-notch impact value, J (min.)	30	25	25

* Limiting section thickness 100 mm

\+ Limiting section thickness 63 mm

Brinell hardness of castings If required by the purchaser and stated on the order, castings heat treated to the requirements of this material specification shall have a Brinell hardness within the following ranges.

Steel A4 152 HB to 207 HB

Steel A5 179 HB to 229 HB

Steel A4 201 HB to 225 HB

2.3 Stainless steels

Worldwide, in industry, in business and in the home, metals called stainless steels are used daily. It is important to understand what these materials are and why they behave the way they do. This is especially true because the word "stainless" is itself somewhat of a misnomer; these materials can stain and can corrode under certain conditions. People need to know why these metals are usually bright and shiny and why they sometimes depart from this expected appearance. In this section, we hope to explain some of these phenomena and provide a better understanding of stainless steels, especially to the non-metallurgist.

The heat treatment may be aimed at the production of a wear-resistant surface or a cutting edge which does not blunt rapidly.

Industries are concerned with integrity of equipment and product purity. To achieve these, stainless steels are often the economical and practical materials of choice for process equipment. However, before intelligent decisions can be made regarding the proper selection from the various types of stainless steel, it is necessary to have an understanding of what stainless steels are. It is important to know what different grades of stainless steel are available, why they perform satisfactorily and why they sometimes do not. In most cases, selection of the proper stainless steel leads to satisfactory performance.

2.3.1 Composition, nomenclature and general properties

Most metals are mixtures of a primary metallic element and one or more intentionally added other elements. These mixtures of elements are called alloys. Stainless steels are alloys, as are brasses (copper + zinc), bronzes (copper + tin), the many aluminum alloys, and many other metallic materials. In general, solid metals and alloys consist of randomly oriented grains that have a well-defined crystalline structure, or lattice, within the grains. In stainless steels, the crystalline structures within the grains have been given names such as ferrite, austenite, martensite, or a mixture of two or more of these. Many of the properties of stainless steels depend upon which crystalline lattice occurs. Examples of these crystal structures are given in Figure 2.15, where the black dots represent atoms and the lines are present to help the structure to be seen.

Ferrite is the basic crystal structure of iron or low-alloy steel at ambient temperature. To understand it, envision a cube with an atom at each of the eight corners and in the geometric center of the cube. This body-centered cubic structure (Figure 2.15(a)) is repeated regularly in three dimensions throughout the grain until it meets a grain of different orientation. At these contacts are areas termed grain boundaries. Grain boundaries consist of many things, including the interface, defects, impurities and grains of other substances. They can be quite complicated and often play an important role in the mechanical properties and corrosion behavior of metals.

Austenite is the crystal form of unalloyed iron in the grains at higher temperatures (> 800 ℃). It is different from ferrite. As in ferrite, there is an atom at each corner of a cube, but instead of one in the geometric center, there is one in the center of each of the six faces of the cube. This face-centered

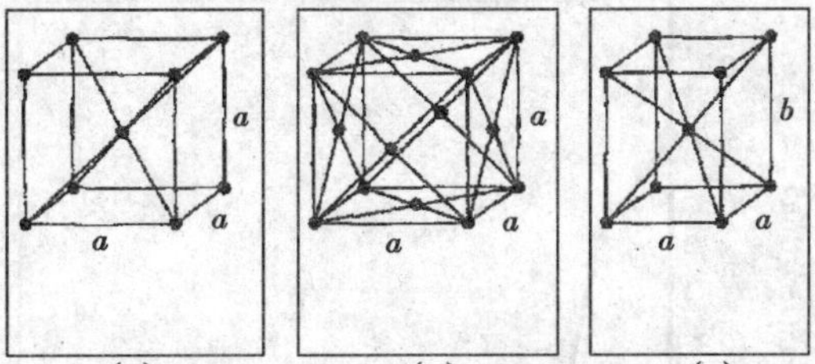

(a) (b) (c)

Figure 2.15 Crystal structures of stainless steels
(a) b.c.c.($a = a = a$) (b) f.c.c.($a = a = a$)
(c) Body-centered structure($a = a \neq b$)

cubic array (Figure 2.15(b)) becomes stable at room temperature if nickel, manganese, nitrogen, or carbon is added, singly or in combination, to iron or iron/chromium alloys. The resulting materials are called austenitic stainless steels. In general, they are easier to shape and bend, more weldable, and less brittle than ferritic alloys.

Martensite is a stable structure at ambient temperature and more similar to ferrite than to austenite. It also has a body-centered structure (Figure 2.15(c)) but one axis of the cube has been elongated to form a tetragonal structure, that is, a crystal having all three axes at right angles and with two equal sides and one unequal. It is produced by heat treatment or cold working cubic crystals of ferrite or austenite. Martensite is the hardest, and strongest of the three crystalline forms, but it is also the least workable. In fact, these alloys are seldom intentionally deformed.

As mentioned previously, alloys are combinations of two or more elements, at least one of which is always a metal. All of the many and varied stainless steels are alloys. They are always iron-chromium alloys, but they often contain other elements, such as molybdenum or nickel. The better known varieties of stainless steel are wrought (hot-rolled or hot-forged after casting into an ingot). There are also cast counterparts that have properties similar to those of most of the wrought grades but that are altered slightly in composition in order to improve casting properties. To define the different materials, the publication Metals and Alloys in the Unified Numbering System lists over 250 types within the broad definition of stainless steels. These are iron base alloys containing more than 11% chromium. Various grades also contain nickel, molybdenum, manganese, nitrogen and other alloying elements. As can be seen in Figure 2.16, chromium's primary effect is to impart corrosion resistance. The diagram shows the influence of chromium on corrosion when it is added to iron or steel. As can be seen, when it reaches 11% — 14%, corrosion is practically negligible in the atmosphere. Nickel in stainless steel promotes austenite stability and reduces the temperature at which austenite can exist. Figure 2.17 illustrates this effect. Above the diagonal line in the diagram, austenite is stable at the indicated temperature; below the line, either ferrite or martensite is the stable crystal structure.

Manganese is similar to nickel when it is added to or substituted for nickel and also increases strength. Molybdenum increases the resistance to localized corrosion phenomena, such as pitting and crevice corrosion. Nitrogen also improves resistance to crevice corrosion, as well as increasing strength and acting as an austenite stabilizer. Elements such as copper and silicon improve corrosion resistance in special environments, and silicon also improves casting properties.

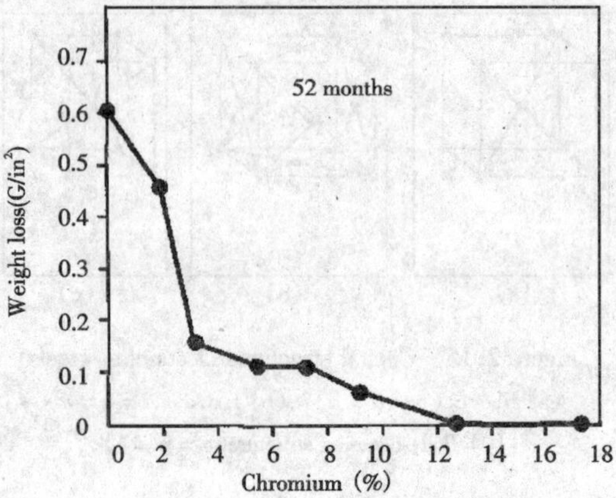

Figure 2.16 The influence of chromium on the atmospheric corrosion of low carbon steel

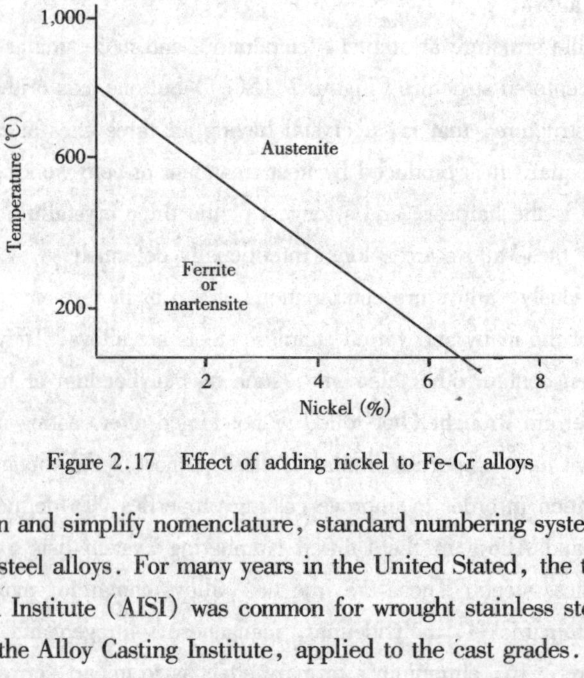

Figure 2.17 Effect of adding nickel to Fe-Cr alloys

To reduce confusion and simplify nomenclature, standard numbering systems have been developed for the various stainless steel alloys. For many years in the United Stated, the three-digit method of the American Iron and Steel Institute (AISI) was common for wrought stainless steels. Another letter and number system, that of the Alloy Casting Institute, applied to the cast grades. These early systems divided stainless steel alloys into groups according to crystal structure. However, many of the newer alloys did not fit into the earlier categories, and it became necessary to have a more complete system. Therefore, these older nomenclatures are now being replaced by the Unified Numbering System (UNS) developed by the Society for Automotive Engineers (SAE) and the American Society for Testing and Materials (ASTM). These groups have developed a six-character notation that assigns a unique designator to metals and alloys in a way that consistently defines a material. For example, the UNS number S30403 replaces AISI304l; the final two digits, 03, indicate the maximum permitted carbon content. In other alloys, the various digits may refer to other parameters; so it cannot always be assumed that the latter numbers mean carbon content. The equivalent cast alloy is J92500, which formerly was ACI CF-3. It should also be noted that the letter before the numbers in the Unified Numbering System pertains to different alloy classes. All letters used will not be defined here, but those of importance will be mentioned. The S refers to heat and corrosion steels (including stainless steels), valve steels and iron-base "superalloys", the J to cast steels (except tool steels), and N to nickel and nickel alloys. The UNS system also provides for classification of many of the newer, more complex alloys that would not fit into the old system, and it covers many types of alloys in addition to the stainless steels.

It should be noted that in other countries different nomenclatures and systems may be used. For example, in Europe the EN system of numbering alloys is in common usage. With this method, S30400 (304) and S31600 (316) are replaced by the numbers 1.4301 and 1.4401, respectively. Most stainless steels have similar designations, some of which are given in format of tables.

Figure 2.18 shows how composition variations have led to many related stainless steels that have evolved from the basic S30400 (304) composition. By altering the composition, as indicated by the arrows and text in the figure, various compositions are produced to meet particular needs. In many cases, this is done by adding or omitting small amounts of other constituents without making major changes in the primary alloy content.

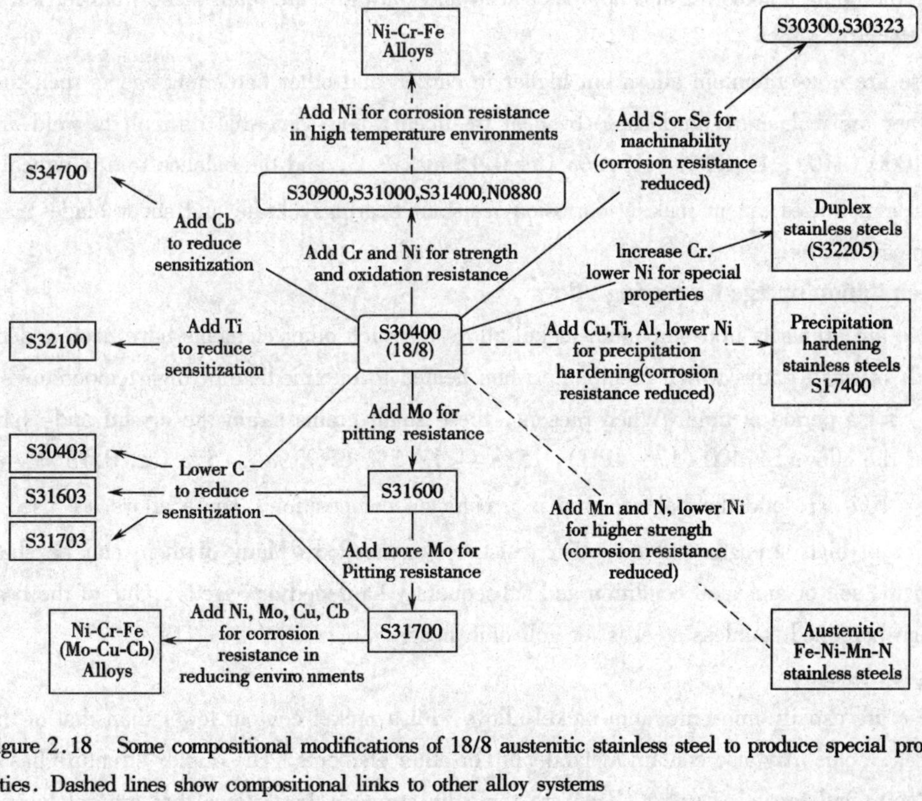

Figure 2.18 Some compositional modifications of 18/8 austenitic stainless steel to produce special properties. Dashed lines show compositional links to other alloy systems

2.3.2 Common stainless steel alloy systems

Austenitic alloy — iron-chromium-nickel and iron-chromium-manganese-nickel alloys

Some of these alloys also contain nitrogen, copper, silicon, and other elements for special purposes. They have an austenitic, i.e., face-centered cubic crystal structure within the grains. To obtain this structure, the austenite/ferrite transition temperature is suppressed by the addition of alloying agents, primarily nickel, but also manganese and nitrogen, so that the resulting austenite is stable at ambient temperature (see Figure 2.17). These alloys are grouped in the 300 and 200 series, respectively, in the old AISI system. They are non-magnetic, unless heavily cold-worked, and hardenable only by cold work. The primary alloy of this type is S30400 (304), 18% — 20% Cr, 8% — 10.5% Ni, and the balance iron. It is commonly referred to as 18-8 stainless steel because of the approximate chromium and nickel contents. Common applications are for an almost endless variety of equipment, including vessels, piping, and tubing used in producing and processing industrial products. Many consumer products such as sinks and wash basins, cooking utensils, pots and pans, and flatware are made from this alloy. When welded fabrication is employed, the low-carbon grade S30403 (304L) is frequently used.

Ferritic alloys — iron-chromium alloys

These have a body-centered cubic crystal structure. They have the corrosion behavior of stainless steels but are sometimes difficult to weld and fabricate. They are magnetic and hardenable only by cold work. An example is S43000(430), 16% — 18% Cr, in the iron base. Less expensive consumer products, including automotive and appliance trim and flatware, are often made from ferritic alloys.

Martensitic alloys

These are iron-chromium alloys but higher in carbon and other hardening agents than the ferritic alloys. They are magnetic, hardenable by heat treatment, and somewhat difficult to weld and fabricate. S41000 (410), 11.5% — 13.5% Cr, 0.15 max% C, and the balance iron is typical of these grades. Common uses are in making corrosion resistant bearings, knife and shear blades, and valve and compressor parts.

Precipitation or age-hardening alloys

These are primarily iron-chromium-nickel alloys to which other elements have been added to form compounds of small grains which precipitate when heated to intermediate or high temperatures (500℃ to 900℃) for a period of time. When present, these small grains strain the crystal and "harden" or strengthen the alloy. S17400 (17-4PH), 15% — 17.5% Cr, 3% — 5% Cu, 0.15% — 0.45% C, 3% — 5% Ni, and the balance iron is a common composition. These alloys are used where a combination of high strength and corrosion resistance is needed. Many of them can be shaped and formed in the soft or annealed condition and subsequently hardened or "aged". One of the best known uses of age-hardened stainless steel is for golf club heads.

Duplex alloys

These are usually iron-chromium-nickel alloys with a nickel content lower than that of the austenitic grades. Some may also contain molybdenum or other elements. The duplex structure has grains of both austenite and ferrite. Duplex alloys are typically stronger than alloys that are solely austenitic, and their corrosion resistance is often at least as good as that of the alloys they replace. Duplex alloys are used in chemical, process, and petroleum industries, especially where better resistance to chloride stress corrosion cracking is required.

2.3.3 Corrosion behavior of stainless steels

The aqueous corrosion of metals is generally considered an electrochemical action. That is, there are alternating sites of differing electrochemical activity on a metal surface. These sites act like the anodes and cathodes in a battery. At the anode, the metal oxidizes (corrodes), reacting with the environment to form rust or some other corrosion product. At the cathode, a reduction reaction such as the reduction of oxygen takes place. This completes the electrochemical cell and corrosion proceeds. In order to prevent corrosion, these cells must be interrupted in some manner.

The unique corrosion resistance of stainless steels is attributed to the existence of a thin, adherent, inactive passive film that covers the surface. This film can conveniently be thought of as chromium oxide, but it also contains small amounts of the other elements in the alloy. Some investigators of the subject consider the film to be something other than an exact oxide, and they may be correct, but it is easier to think of the film as an oxide. Many people think stainless steel must be given a "passivating" treatment for this film to form properly. This is not true; if the surface is clean and free of contamination, the film forms instantaneously on exposure to air, aerated water, nitric acid, or other oxidizing media. It is extremely durable and reforms spontaneously.

Because of this protective film, stainless steels do not corrode as carbon or low alloy steels or cast iron do. These materials "rust" or corrode uniformly through constantly changing anodes and cathodes on the surface. However, except in solutions such as hydrochloric acid, this general corrosion or uniform attack practically never occurs on stainless steels. The terms "corrosion rate" and "corrosion allowance" are usually meaningless when applied to stainless steels. While factors such as chemical environment, pH, temperature, equipment design, fabrication methods, surface finish, contamination, and maintenance procedures can affect the corrosion of stainless steels, they usually cause only some form of localized corrosion. To explain this further, the various types of localized corrosion — pitting, crevice corrosion, intergranular corrosion, stress corrosion cracking, and galvanic corrosion — will be considered separately.

Pitting and crevice corrosion

Because pitting and crevice corrosion are very similar and the factors that affect their occurrence are essentially the same, these two phenomena will be considered together. Pitting (Figure 2.19) is highly localized corrosion at individual sites on the surface of the metal. The figure also shows that pits vary in size, shape, and morphology. Some pits are broad and not very deep while some penetrate quite deeply and others may undercut the passive film and spread out beneath it. Crevice corrosion (Figure 2.20) is the attack that occurs at the interface between the corroding metal and another substance, usually one that is not electrically conductive. The corrosion usually spreads into the crevice beyond the point of contact. Both types of corrosion happen on stainless steel in certain media, especially those containing chlorides. Pitting can occur because of minor discontinuities in the passive film, inclusions or defects in the stainless steel, or dirt and contamination on the surface. Examples of common crevices are joints with gaskets, at points where scale or hard biofouling attaches and in places where materials overlap. Because the area of the attack is very small in comparison to the overall area of the metal surface, corrosion can be very intense and rapid at the site of attack. The most important single fact in the initiation of crevice corrosion is the presence of chloride ions, although higher environmental temperatures, oxygen or easily reducible ions such as ferric ions, and acid pH values can also have deleterious effects. Pitting is less apt to occur in aqueous solutions moving at moderate to high velocities than in stagnant ones.

Figure 2.19 Pitting corrosion

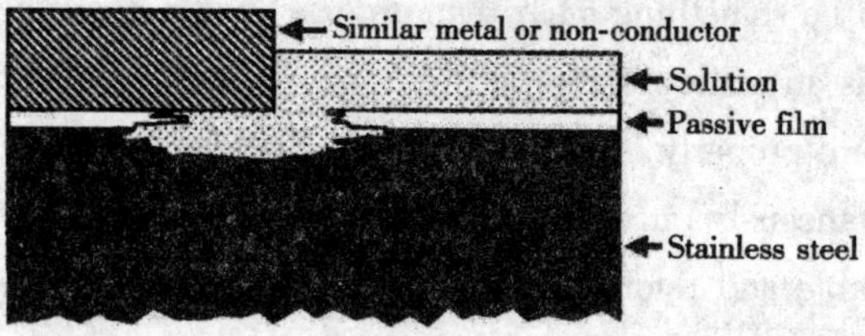

Figure 2.20 Crevice corrosion

Although we have previously said that pitting and crevice corrosion are essentially the same, some differences should be mentioned. Crevice corrosion can occur in environments that normally do not cause pitting in boldly exposed sheet or plate, particularly in tight stationary crevices in slow-moving solutions.

If the environment cannot be controlled, by reducing acidity, or chloride content or by increasing solution velocity, more highly alloyed grades may be used to control pitting and crevice corrosion. This is usually done by adjusting chromium and nickel content and adding more molybdenum (and, to a lesser extent, nitrogen) to the alloys. The pitting resistance of a common material such as S30400 with no added molybdenum can be markedly improved in this way. Alloys such as S31600 (2.3% Mo), S31700 (3% — 4% Mo), N08904 (4% — 5% Mo), and the 6% — 7% Mo alloys have increasing pitting and crevice corrosion resistance with increasing molybdenum content. Good design and fabrication techniques that produce smooth, clean surfaces, rounded corners, and "drain away" designs also help resist pitting and crevice corrosion.

Intergranular corrosion

If an austenitic stainless steel of normal carbon content (0.03% — 0.08% C) is heated in the temperature range from 425 ℃ (800 ℉) to 815 ℃ (1,500 ℉), chromium carbides are precipitated at grain boundaries and the structure is said to be "sensitized". The chromium-depleted zone around each grain is more susceptible to attack in some media, particularly acids. Exposure to this critical temperature range can result from improper annealing, stress relieving, or heating during forming and welding. Figure 2.21 is a representation of what can happen. As is seen, corrosion has proceeded from the surface down the grain boundaries to the extent that the grains can become detached and the surface is sometimes said to have "sugared". Other than heat treatment, there are usually two solutions to sensitization: use of a low-carbon alloy such as S30403 (304L) or use of an alloy containing, or "stabilized" with, titanium, S32100 (321), or columbium, S34700 (347). In the first case, there is insufficient carbon in the alloy to form large amounts of chromium carbides and thus reduce chromium in the grain boundaries. In the latter case, the carbon is precombined with titanium or columbium and is therefore not available to the chromium. The titanium and columbium carbides are dispersed in the matrix of the grains and not localized at grain boundaries to promote intergranular corrosion. In recent years, as the AOD and VOD processes have become more successful and low carbon alloys easier to produce, the low carbon grades of stainless steel have largely supplanted the stabilized alloys for welded fabrication. Because of their ease of production, they are also replacing the standard carbon grades for many applications.

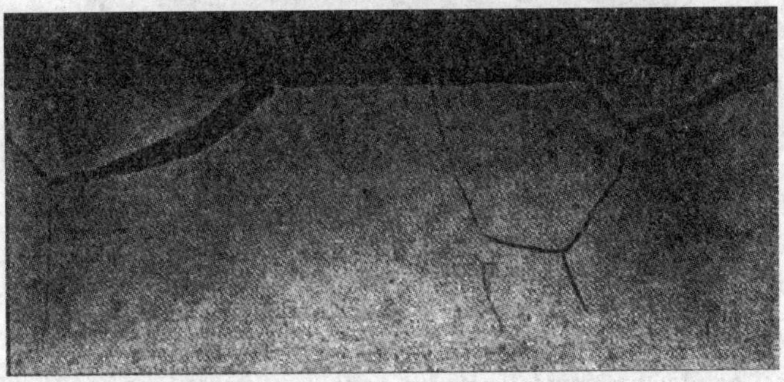

Figure 2.21 Intergranular attack in a sensitized austenitic alloy produced by exposure to a boiling sulfuric acid-ferric sulfate solution. Prolonged exposure causes grains to detach from surface(× 100)

Stress corrosion cracking

The phenomenon of stress corrosion cracking (Figure 2.22) of austenitic stainless steels in chloride-containing environments is not unique to stainless steels. Many types of alloys are susceptible to similar effects in different media, such as brass alloys in ammoniacal environments and carbon and alloy steels, including stainless steels, in strongly alkaline solutions. Chloride stress corrosion cracking, the most common form of environmentally induced cracking in austenitic stainless steels, requires the presence of chloride ions, tensile stresses, and elevated temperature. If these are moderate to low, oxygen is also required for stress corrosion cracking to occur. The necessary tensile stresses are almost always residual rather than applied. It is not the load put on a stainless steel vessel that leads to cracking, but how it is formed and welded. In properly annealed material, the cracking is characteristically transgranular (across the grains). In poorly heat-treated and in weld heat-affected zones where carbides have precipitated at grain boundaries, the cracking is intergranular (at the grain boundaries). Minimum levels of chloride content, temperature, and stress are not known, because these variables are interrelated. The phenomenon is usually controlled by proper alloy selection, although altering the environment and reducing residual stresses can sometimes be effective. In general, ferritic and duplex stainless steels have more resistance to chloride stress corrosion cracking and are often substituted. Austenitic iron-nickel-chromium alloys also have increased resistance at nickel contents above 20%. In fact, some of the 6% — 7% Mo alloys with 17% — 23% Cr and 17% — 26% Ni have good resistance to chloride stress corrosion cracking. However, virtual immunity is probably found in austenitic alloys only when nickel levels are above 35%.

Galvanic corrosion

Galvanic corrosion, or dissimilar metal corrosion, is usually not a problem for stainless steels but

Figure 2.22 Transgranular chloride stress corrosion cracking of an austenitic stainless steel

can affect other metals in contact with them. For galvanic corrosion to take place, two or more metals of different electrochemical activity need to be in intimate contact in an electrolyte solution. An abbreviated galvanic series, or electrochemical activity series, of materials in seawater is given in Table 2.10. The standard hydrogen electrode is used as a reference against which electrochemical activity of a material is measured: The activity of hydrogen is set at zero and other materials are measured as more active (–) or more passive (+) with regard to it. In this table, the more negative or active metals (at the top of the table) will corrode preferentially to any less active metal to which they are electrically coupled. If the surface area of the active component is small in relation to that of the other member of the couple, the corrosion rate can be very high. Such would be the case if carbon steel bolts or rivets were used to connect stainless steels sheet or plate. Sometimes, when the film is disrupted, stainless steel can become the active metal (as shown in Table 2.10) and corrode in an active manner. Also, once pitting and crevice corrosion begin, these forms can be considered galvanic corrosion. In both cases, the result is a small active area (the pits or the crevices) surrounded by a large area of film-protected, inactive stainless steel. Most galvanic corrosion problems can be avoided by proper design or electrical insulation.

Table 2.10 Galvanic series of some metals and alloys in seawater

	Metal or Alloy	Potential vs. SHE[1]
Active (anodic)	Magnesium	–1.49
	Zinc	–0.81
	Cadmium	–0.64
	Aluminum	–0.61
	Steel	–0.38
	S30400 Stainless Steel (active)	–0.36
	Lead	–0.32
	Tin	–0.27
	Admiralty Brass	–0.12
	Hydrogen	0.00
	Copper	+0.02
	Nickel	+0.10

Table 2.10 Galvanic series of some metals and alloys in seawater (Continued)

	Metal or Alloy	Potential vs. SHE[1]
	N04400 (Monel Ni/Cu Alloy)	+0.13
	Titanium	+0.14
	S30400 Stainless Steel (passive)	+0.15
	Silver	+0.16
	Graphite	+0.49
	Platinum	+0.50
Passive (cathodic)	Gold	+0.50

[1] SHE: Standard Hydrogen Electrode

2.4 Nonferrous alloys

Steels and other ferrous alloys are consumed in exceedingly large quantities because they have such a wide range of mechanical properties, may be fabricated with relative ease, and are economical to produce. However, they have some distinct limitations, chiefly: ① a relatively high density, ② a comparatively low electrical conductivity, and ③ an inherent susceptibility to corrosion in some common environments. Thus, for many applications it is advantageous or even necessary to utilize other alloys having more suitable property combinations. Alloy systems are classified either according to the base metal or according to some specific characteristic that a group of alloys share. This section discussed the following metal and alloy systems: copper, aluminum, magnesium, and titanium alloys, the refractory metals, the superalloys, the noble metals, and miscellaneous alloys, including those that have nickel, lead, tin, zirconium, and zinc as base metals.

On occasion, a distinction is made between cast and wrought alloys. Alloys that are so brittle that forming or shaping by appreciable deformation is not possible ordinarily are cast; these are classified as cast alloys. On the other hand, those that are amenable to mechanical deformation are termed wrought alloys.

In addition, the heat treatability of an alloy system is mentioned frequently. "Heat treatable" designates an alloy whose mechanical strength is improved by precipitation hardening or a martensitic transformation (normally the former), both of which involve specific heat-treating procedures.

2.4.1 Copper and its alloys

Copper and copper-based alloys, possessing a desirable combination of physical properties, have been utilized in quite a variety of applications since antiquity. Unalloyed copper is so soft and ductile that it is difficult to machine; also, it has an almost unlimited capacity to be cold worked. Furthermore, it is highly resistant to corrosion in diverse environments including the ambient atmosphere, seawater, and some industrial chemicals. The mechanical and corrosion-resistance properties of copper may be improved by alloying. Most copper alloys cannot be hardened or strengthened by heat-treating procedures; consequently, cold working and/or solid-solution alloying must be utilized to improved these mechanical properties.

The most common copper alloys are the brasses for which zinc, as a substitutional impurity, is the predominant alloying element. As may be observed from the copper-zinc phase diagram, the α phase is stable for concentrations up to approximately 35 wt% Zn. This phase has an FCC crystal structure, and α brasses are relatively soft, ductile, and easily cold worked. Brass alloys having a higher zinc content contain both α and β′ phases at room temperature. The β′ phase has an ordered BCC crystal structure and is harder and stronger than the α phase; consequently, α + β′ alloys are generally hot worked.

Some of the common brasses are yellow, naval, and cartridge brass, muntz metal, and gilding metal. The compositions, properties, and typical uses of several of these alloys are listed in Table 2.11. Some of the common uses for brass alloys include costume jewelry, cartridge casings, automotive radiators, musical instruments, electronic packaging, and coins.

The bronzes are alloys of copper and several other elements, including tin, aluminum silicon, and nickel. These alloys are somewhat stronger than the brasses, yet they still have a high degree of corrosion resistance. Table 2.11 contains several of the bronze alloys, their compositions, properties, and application. Generally they are utilized when, in addition to corrosion resistance, good tensile properties are required.

The most common precipitation hardenable copper alloys are the beryllium coppers. They possess a remarkable combination of properties: tensile strengths as high as 1,400MPa (200,000 psi), excellent electrical and corrosion properties, and wear resistance when properly lubricated; they may be cast, hot worked, or cold worked. High strengths are attained by precipitation-hardening heat treatments. These alloys are costly because of the beryllium additions, which range between 1.0 and 2.5wt%. Applications include jet aircraft landing gear bearings and bushings, springs, and surgical and dental instruments. One of these alloys (C17200) is included in Table 2.11.

Table 2.11 Compositions, mechanical properties, and typical applications for eight copper alloys

Alloy Name	UNS Number	Composition (wt%)*	Condition	Mechanical Properties: Tensile Strength [MPa (ksi)]	Yield Strength [MPa (ksi)]	Ductility [% El in 50 mm (2 in.)]	Typical Applications
				Wrought Alloys			
Electrolytic tough pitch	C 11000	0.04 O	Annealed	220(32)	69(10)	45	Electrical wire, rivets, screening, gasket, pans, nails, roofing
Beryllium copper	C 17200	1.9Be, 0.20Co	Precipitation hardened	1,140 — 1,310 (165 — 190)	690 — 860 (100 — 125)	4 — 10	Springs, bellows, firing pins, bushings, valves, diaphragms
Cartridge brass	C 26000	30Zn	Annealed	325(47)	130 (19)	68	Automotive radiator cores, ammunition components, lamp fixtures, flashlight shells, kickplates
			Cold-worked (H04 hard)	560(81)	515(75)	8	

Table 2.11 Compositions, mechanical properties, and typical applications for eight copper alloys (Continued)

Alloy Name	UNS Number	Composition (wt%)*	Condition	Mechanical Properties			Typical Applications
				Tensile Strength [MPa (ksi)]	Yield Strength [MPa (ksi)]	Ductility [% El in 50 mm (2 in.)]	
Phosphor bronze, 5% A	C 51000	5An, 0.2P	Annealed	325(47)	130(19)	64	Bellows, clutch disks, diaphragms, fuse clips, springs, welding rods
			Cold-worked (H02 hard)	515(75)	485(70)	15	
Cast Alloys							
Leaded yellow brass	C 85400	29Zn, 3Pb, 1Sn	As cast	234(34)	83(12)	35	Furniture hardware, radiator fittings, light fixtures, battery champs
Tin bronze	C 90500	10Sn, 2 Zn	As cast	310(45)	152(22)	25	Bearings, bushings, piston rings, steam fittings, gears
Aluminum bronze	C 95400	4Fe, 11Al	As cast	586(85)	241(35)	18	Bearing, gears, bushings, valve seats and guards

* The balance of the composition is copper.

2.4.2 Aluminum and its alloys

Aluminum and its alloys are characterized by a relatively low density (2.7 g/cm^3 as compared to 7.9 g/cm^3 for steel), high electrical and thermal conductivities, and a resistance to corrosion in some common environments, including the ambient atmosphere. Many of these alloys are easily formed by virtue of high ductility; this is evidenced by the thin aluminum foil sheet into which the relatively pure material may be rolled. Since aluminum has an FCC crystal structure, its ductility is retained even at very low temperatures. The chief limitation of aluminum is its low melting temperature (660 ℃ (1 220 ℉)), which restricts the maximum temperature at which it can be used.

The mechanical strength of aluminum may be enhanced by cold work and by alloying; however, both processes tend to diminish resistance to corrosion. Principal alloying elements include copper, magnesium, silicon, manganese, and zinc. Non-heat-treatable alloys consist of a single phase, for which an increase in strength is achieved by solid solution strengthening. Others are rendered heat treatable (capable of being precipitation hardened) as a result of alloying. In several of these alloys precipitation hardening is due to the precipitation of two elements other than aluminum, to form an intermetallic compound such as $MgZn_2$.

Generally, aluminum alloys are classified as either cast or wrought. Composition for both types is designated by a four-digit number that indicates the principal impurities, and in some cases, the purity level. For cast alloys, a decimal point is located between the last two digits. After these digits is a hyphen and the basic temper designation — a letter and possibly a one-or three-digit number, which indicates the mechanical and/or heat treatment to which the alloy had been subjected. For example, F,

H, and O represent, respectively, the as-fabricated, strain-hardened, and annealed states; T3 means that the alloy was solution heat treated, cold worked, and then naturally aged (age hardened). A solution heat treatment followed by artificial aging is indicated by T6. The compositions, properties, and applications of several wrought and cast alloys are contained in Table 2.12. Some of the more common applications of aluminum alloys include aircraft structural parts, beverage cans, bus bodies, and automotive parts (engine blocks, pistons, and manifolds).

Recent attention has been given to alloys of aluminum and other low-density metals (e.g., Mg and Ti) as engineering materials for transportation, to effect reductions in fuel consumption. An important characteristic of these materials is specific strength, which is quantified by the tensile strength-specific gravity ratio. Even though an alloy of one these metals may have a tensile strength that is inferior to a more dense material (such as steel), on a weight basis it will be able to sustain a larger load.

A generation of new aluminum-lithium alloys have been developed recently for use by the aircraft and aerospace industries. These materials have relatively low densities (between about 2.5 and 2.6 g/cm^3), high specific moduli (elastic modulus-specific gravity ratios), and excellent fatigue and low-temperature toughness properties. Furthermore, some of them may be precipitation hardened. However, these materials are more costly to manufacture than the conventional aluminum alloys because special processing techniques are required as a result of lithium's chemical reactivity.

Table 2.12 Compositions, mechanical properties, and typical applications for several common aluminum alloys

Alloy Name	UNS Number	Composition (wt%)*	Condition	Mechanical Properties: Tensile Strength [MPa (ksi)]	Yield Strength [MPa (ksi)]	Ductility [% El in 50 mm (2 in.)]	Typical Applications /Characterstics
Wrought Nonheat-Treatable Alloys							
1100	A91100	0.12Cu	Annealed (O)	90(13)	35(5)	35 — 45	Food/chemical handling & storage equipment, heat exchangers, light reflectors
3003	A93003	0.12Cu, 1.2Mn, 0.1Zn	Annealed (O)	110(16)	40(6)	30 — 40	Cooking utensils, pressure vessels and piping
5052	A95052	2.5Mg, 0.25Cr	Strain hardened (H32)	230(33)	195(28)	12.18	Aircraft fuel & oil lines, fuel tanks, appliances, rivets, and wire
Wrought, Heat-Treatable Alloys							
2024	A92024	4.4Cu, 1.5Mg, 0.6Mn	Heat treated (T4)	470(68)	325(47)	20	Aircraft structures, rivets, truck wheels, screw machine products
6061	A96061	1.0Mg, 0.6Si, 0.30Cu, 0.20Cr	Heat treated (T4)	240(35)	145(21)	22.25	Trucks, canoes, railroad cars, furniture, pipelines
7075	A97075	5.6Zn, 2.5Mg, 1.6Cu, 0.23Cr	Heat treated (T6)	570(83)	505(73)	11	Aircraft structural parts and other highly stressed applications

Table 2.12 Compositions, mechanical properties, and typical applications for several common aluminum alloys (Continued)

Alloy Name	UNS Number	Composition (wt%)*	Condition	Mechanical Properties: Tensile Strength [MPa (ksi)]	Yield Strength [MPa (ksi)]	Ductility [% El in 50 mm (2 in.)]	Typical Applications /Characterstics
Cast, Heat-Treatable Alloys							
295.0	A02950	4.5Cu, 1.1Si	Heat treated (T4)	221(32)	110(16)	8.5	Flywheel and rear-axle housings, bus and aircraft wheels, crankcases
356.0	A03560	7.0Si, 0.3Mg	Heat treated (T6)	228(33)	164(24)	3.5	Aircraft pump parts, automotive transmission cases, water-cooled cylinder blocks
Aluminum-Lithium Alloys							
2090	–	2.7Cu, 0.25Mg, 2.25Li, 0.12 Zr	Heat treated, cold worked (T83)	455(66)	455(66)	5	Aircraft structures and cryogenic tank-age structures
8090	–	1.3Cu, 0.95 Mg, 2.0Li, 0.1Zr	Heat treated, cold worked (T651)	465(67)	360(52)	–	Aircraft structures that must be highly damage tolerant

* The balance of the composition is aluminum.

2.4.3 Magnesium and its alloys

Perhaps the most outstanding characteristic of magnesium is its density, 1.7 g/cm^3, which is the lowest alloy of the structural metals; therefore, its alloys are used where light weight is an important consideration (e.g., in aircraft components). Magnesium has an HCP crystal structure, is relatively soft, and has a low elastic modulus: 45 GPa (6.5×10^6 psi). At room temperature magnesium and its alloys are difficult to deform; in fact, only small degrees of cold work may be imposed without annealing. Consequently, most fabrication is by casting or hot working at temperatures between 200 and 350 ℃ (400 and 650 ℉). Chemically, magnesium alloys are relatively unstable and especially susceptible to corrosion in marine environments. On the other hand, corrosion or oxidation resistance is reasonably good in the normal atmosphere; it is believed that this behavior is due to impurities rather than being an inherent characteristic of Mg alloys. Fine magnesium powder ignites easily when heated in air; consequently, care should be exercised when handling it in this state.

These alloys are also classified as either cast or wrought, and some of them are heat treatable. Aluminum, zinc, manganese, and some of the rare earths are the major alloying elements. A composition-temper designation scheme similar to that for aluminum alloys is also used. Table 2.13 lists several common magnesium alloys, their compositions, properties, and applications. These alloys are used in aircraft and missile applications, as well as in luggage. Furthermore, in the last several years the demand for magnesium alloys has increased dramatically in a host of different industries. For many applications, magnesium alloys have replaced engineering plastics that have comparable densities inasmuch as the magnesium materials are stiffer, more recyclable, and less costly to produce. For example, magnesium is now employed in a variety of hand-held devices (e.g., chain saws, power tools, hedge clippers), in automobiles (e.g., steering wheels and columns, seat frames, transmission cases), and in audio-video-computer communications equipment (e.g., laptop computers, camcorders, TV sets, cellular telephones).

Table 2.13 Compositions, mechanical properties, and typical applications for six common magnesium alloys

Alloy Name	UNS Number	Composition (wt%)*	Condition	Mechanical Properties			Typical Applications
				Tensile Strength [MPa (ksi)]	Yield Strength [MPa (ksi)]	Ductility [% El in 50 mm (2 in.)]	
				Wrought Alloys			
AZ31B	M11311	3.1 Al, 1.0 Zn, 0.2 Mn	As extruded	262 (38)	200 (29)	15	Structures and tubing, cathodic protection
HK31A	M13310	3.0 Th, 0.6 Zr	Strain hardened, partially annealed	255 (37)	200 (29)	9	High strength to 315 ℃ (600 ℉)
ZK 60A	M16600	5.5 Zn, 0.45 Zr	Artificially aged	350 (51)	285 (41)	11	Forgings of maximum strength for aircraft
				Cast Alloys			
AZ91D	M11910	9.0 Al, 0.15 Mn, 0.7 Zn	As cast	230 (33)	150 (22)	3	Die-cast parts for automobiles, huggage, and electronic devices
AM60A	M10600	6.0 Al, 0.13 Mn	As cast	220 (32)	130 (19)	6	Automotive wheels
AS41A	M10410	4.3 Al, 1.0 Si, 0.35 Mn	As cast	210 (31)	140 (20)	6	Die castings requiring good creep resistance

* The balance of the composition is magnesium.

2.4.4 Titanium and its alloys

Titanium and its alloys are relatively new engineering materials that possess an extraordinary combination of properties. The pure metal has a relatively low density (4.5 g/cm^3), a high melting point [1,668 ℃ (3,035 ℉)], and an elastic modulus of 107 GPa (15.5×10^6 psi). Titanium alloys are extremely strong, room temperature tensile strengths as high as 1,400 MPa (200,000 psi) are attainable, yielding remarkable specific strengths. Furthermore, the alloys are highly ductile and easily forged and machined.

The major limitation of titanium is its chemical reactivity with other materials at elevated temperatures. This property has necessitated the development of non-conventional refining, melting, and casting techniques; consequently, titanium alloys are quite expensive. In spite of this high temperature reactivity, the corrosion resistance of titanium alloys at normal temperatures is unusually high; they are virtually immune to air, marine, and a variety of industrial environments. Table 2.14 presents several titanium alloys along with their typical properties and applications. They are commonly utilized in airplane structures, space vehicles, surgical implants, and in the petroleum and chemical industries.

Table 2.14 Compositions, mechanical properties, and typical applications for six common titanium alloys

Alloy Name	UNS Number	Composition (wt%)	Condition	Average Mechanical Properties: Tensile Strength [MPa (ksi)]	Yield Strength [MPa (ksi)]	Ductility [% El in 50 mm (2 in.)]	Typical Applications
Commercially pure	Unalloyed (R50500)	99.1Ti	Annealed	484 (70)	414 (60)	25	Jet engine shrouds, cases and airframe skins, corrosion-resistant equipment for marine and chemical processing industries
αβ	Ti-5Al-2.5Sn (R54520)	5Al, 2.5Sn, balance Ti	Annealed	826 (120)	784 (114)	16	Gas turbine engine casings and rings; chemical processing equipment requiring strength to temperatures of 480℃ (900 ℉)
Near α	Ti-8Al-1Mo-1V (R54810)	8Al, 1Mo, 1 V, balance	Annealed (duplex)	950 (138)	890 (129)	15	Forgings for jet engine components (compressor disks, plates, and hubs)
α-β	Ti-6Al-4V (R56400)	6Al, 4V, balance Ti	Annealed	947 (137)	877 (127)	14	High-strength prosthetic implants, chemical-processing equipment, airframe structural components
α-β	Ti-6Al-6V-2Sn (R56620)	6Al, 2Sn, 6 V, 0.75Cu, balance Ti	Annealed	1050 (153)	985 (143)	14	Rocket engine case airframe applications and high-strength airframe structures
β	Ti-10V-2Fe-3Al	10V, 2Fe, 3Al, balance Ti	Solution + aging	1223(178)	1150(167)	10	Best combination of high strength and toughness of any commercial titanium alloy; used for applications requiring uniformity of tensile properties at surface and center location; high-strength air frame components

2.4.5 Refractory metals

Metals that have extremely high melting temperatures are classified as the refractory metals. Included in this group are niobium (Nb), molybdenum (Mo), tungsten (W), and tantalum (Ta). Melting temperatures range between 2,468 ℃ (4,474 ℉) for niobium and 3,410 ℃ (6,170 ℉), the highest melting temperature of any metal, for tungsten. Interatomic bonding in these metals is extremely strong, which accounts for the melting temperatures, and, in addition, large elastic moduli and high strengths and hardnesses, at ambient as well as elevated temperatures. The applications of these metals are varied. For example, tantalum and molybdenum are alloyed with stainless steel to improve its corrosion resistance. Molybdenum alloys are utilized for extrusion dies and structural parts in space vehicles; incandescent light filaments, X-ray tubes, and welding electrodes employ tungsten alloys. Tantalum is immune to chemical attack by virtually all environments at temperatures below 150℃, and is frequently used in applications requiring such corrosion-resistant material.

2.4.6 Superalloys

The superalloys have superlative combinations of properties. Most are used in aircraft turbine components, which must withstand exposure to severely oxidizing environments and high temperatures for reasonable time period. Mechanical integrity under these conditions is critical; in this regard, density is an important consideration because centrifugal stresses are diminished in rotating members when the density is reduced. These materials are classified according to the predominant metal in the alloy, which may be cobalt, nickel, or iron. Other alloying elements include the refractory metals (Nb, Mo, W, Ta), chromium, and titanium. In addition to turbine applications, these alloys are utilized in nuclear reactors and petrochemical equipments.

2.4.7 Noble metals

The noble or precious metals are a group of eight elements that have some physical characteristics in common. They are expensive (precious) and are superior or notable (noble) in properties — i.e., characteristically soft, ductile, and oxidation resistant. The noble metals are silver, gold, platinum, palladium, rhodium, ruthenium, iridium, and osmium; the first three are most common and are used extensively in jewelry. Silver and gold may be strengthened by solid-solution alloying with copper; sterling silver and is a silver-copper alloy containing approximately 7.5 wt% Cu. Alloys of both silver and gold are employed as dental restoration materials; also, some integrated circuit electrical contacts are of gold. Platinum is used for chemical laboratory equipments, as a catalyst (especially in the manufacture of gasoline), and in thermocouples to measure elevated temperatures.

2.4.8 Miscellaneous nonferrous alloys

The discussion above covers the vast majority of nonferrous alloys; however, a number of others are found in a variety of engineering applications, and a brief exposure of these is worthwhile.

Nickel and its alloys are highly resistant to corrosion in many environments, especially those that are basic (alkaline). Nickel is often coated or plated on some metals that are susceptible to corrosion as a protective measure. Monel, a nickel-based alloy containing approximately 65 wt% Ni and 28% Cu (the balance iron), has very high strength and is extremely corrosion resistant; it is used in pumps, valves, and other components, that are in contact with some acid and petroleum solutions. As already mentioned, nickel is one of the principal alloying elements in stainless steels, and one of the major constituents in the superalloys.

Lead, tin, and their alloys find some use as engineering materials. Both are mechanically soft and weak, have low melting temperatures, are quite resistant to many corrosion environments, and have recrystallization temperatures below room temperature. Many common solders are lead-tin alloys, which have low melting temperatures. Applications for lead and its alloys include X-ray shields and storage batteries. The primary use of tin is as a vary thin coating on the inside of plain carbon steel cans (tin cans) that are used for food containers; this coating inhibits chemical reactions between the steel and the food products.

Unalloyed zinc also is a relatively soft metal having a low melting temperature and a subambient recrystallization temperature. Chemically, it is reactive in a number of common environments and, therefore, susceptible to corrosion. Galvanized steel is just plain carbon steel that has been coated with a thin zinc layer; the zinc preferentially corrodes and protects the steel. Typical applications of galvanized steel are familiar (sheet metal, fences, screen, screws, etc.). Common applications of zinc alloys include padlocks, automotive parts (door handles and grilles), and office equipments.

Although zirconium is relatively abundant in the earth's curst, it was not until quite recent times that commercial refining techniques were developed. Zirconium and its alloys are ductile and have other mechanical characteristics that are comparable to those of titanium alloys and the austenitic stainless steels. However, the primary asset of these alloys is their resistance to corrosion in a host of corrosive media, including superheated heater. Furthermore, zirconium is transparent to thermal neutrons, so that its alloys have been used as cladding for uranium fuel in water-cooled nuclear reactors. In terms of cost, these alloys are also often the materials of choice for heat exchangers, reactor vessels, and piping systems for the chemical-processing and nuclear industries. They are also used in incendiary ordnance and in sealing devices for vacuum tubes.

Keywords

austenite	annealing	cementite
crevice corrosion	cracking passivate	dislocation
ferrite	grain boundary	galvanic corrosion
heat treatment	hardening	impact strength
lattice	mechanical property	martensite
normalizing	pearlite	pitting
quenching	stainless steel	superalloys
stress corrosion	tensile strength	tempering

Problems

1. (a) List the four classifications of steels. (b) For each, briefly describe the properties and typical applications.

2. What is the function of alloying elements in common steels and stainless steels?

3. If a heat treatment is wanted as the preparation for mechanical working, which one is suitable, the full annealing, the normalizing or the hardening? And why?

4. Give the distinctive features, limitations, and applications of the following alloy groups: titanium alloys, refractory metals, superalloys, and noble metals.

5. For the following pairs of alloys that are coupled in seawater, predict the possibility of corrosion; if corrosion is probable, note which alloy will corrode.

(a) Aluminum and magnesium.

(b) Zinc and a low-carbon steel.

(c) Brass (60Cu-40Zn) and Monel (70 Ni-30Cu).

(d) Titanium and 304 stainless steel.

(e) Cast iron and 316 stainless steel.

6. Briefly explain why ferritic and austenitic stainless steels are not heat treatable.

7. Why is the nickel content in a duplex alloys generally lower than that of the austenitic grades?

8. Briefly explain why the hardness of tempered martensite diminishes with tempering time (at constant temperature) and with increasing temperature (at constant tempering time).

9. In your own words describe the following heat treatment procedures for steels and, for each, the intended final microstructure: full annealing, normalizing, quenching, and tempering.

Selected from

[1] L. M. Grourd. An Introduction to Engineering Materials. Edward Arnold ltd., 1982.

[2] R. A. Covert and Arthur H. Tuthill, 2000,20(7):506 — 517.

[3] ASM Handbook. Vol.2, Properties and Selection: Nonferrous Alloys and Special-Purpose Materials. ASM International, 1990.

[4] W. D. Callister. Fundamentals of Materials Science and Engineering. 5th Edition. John Wiley & Sons, 2001

3 Ceramic Materials

3.1 Introduction to ceramics

3.1.1 Definition of ceramics

The word ceramic can be traced back to the Greek term keramos, meaning "a potter" or "pottery". Keramos in turn is related to an older Sanskrit root meaning "to burn". Thus the early Greeks used the term to mean "burned stuff" or "burned earth" when referring to products obtained through the action of fire upon earthy materials. The term ceramic's meaning has evolved a great deal from its Greek root keramos, meaning pottery (or potter's clay). Ceramic materials can now be broadly considered to be all inorganic non-metallic materials. They are typically crystalline in nature and are compounds formed between metallic and nonmetallic elements such as aluminum and oxygen (alumina — Al_2O_3), calcium and oxygen (calcia — CaO), and silicon and nitrogen (silicon nitride — Si_3N_4). But in fact, they can be crystalline, non-crystalline, or a mixture of both. Moreover, it is more useful to classify them as inorganic non-metallic materials that acquire mechanical strength through a sintering process.

3.1.2 Classification of ceramics

The broad categories or segments that make up the ceramic industry can be classified as follows:

(a) structural clay products

(b) whitewares

(c) refractories

(d) glasses

(e) abrasives

(f) cements

(g) advanced ceramics

Most of the above segments can be further broken down into more specific product classifications as seen on the table below (Table 3.1). In 1974, the U.S. market for the ceramic industry was estimated at $ 20 million. Today, the U.S. market is estimated to be over $ 35 billion.

Table 3.1 Industries of Ceramic Segments

Industry Segment	Common Examples
Structural clay products	Brick, sewer pipe, roofing tile, clay floor and wall tile (i.e., quarry tile), flue linings
Whitewares	Dinnerware, floor and wall tile, sanitaryware, electrical porcelain, decorative ceramics
Refractories	Brick and monolithic products are used in iron and steel, non-ferrous metals, glass, cements, ceramics, energy conversion, petroleum, and chemicals industries
Glasses	Flat glass (windows), container glass (bottles), pressed and blown glass (dinnerware), glass fibers (home insulation), and advanced/specialty glass (optical fibers)
Abrasives	Natural (garnet, diamond, etc.) and synthetic (silicon carbide, diamond, fused alumina, etc.) abrasives are used for grinding, cutting, polishing, lapping, or pressure blasting of materials
Cements	Used to produce concrete roads, bridges, buildings, dams, and the like
Advanced ceramics	
Structural	Wear parts, bioceramics, cutting tools, and engine components
Electrical	Capacitors, insulators, substrates, integrated circuit packages, piezoelectrics, magnets and superconductors
Coatings	Engine components, cutting tools, and industrial wear parts
Chemical and environmental	Filters, membranes, catalysts, and catalyst supports

3.1.3 Overview of ceramic and glass manufacturing

Ceramics are typically produced by the application of heat upon processed clays and other natural raw materials to form a rigid product. Ceramic products that use naturally occurring rocks and minerals as starting materials must undergo special processing in order to control purity, particle size, particle size distribution, and heterogeneity. These attributes play a big role in the final properties of the finished ceramic. Chemically prepared powders also are used as starting materials for some ceramic products. These synthetic materials can be controlled to produce powders with precise chemical compositions and particle size.

The next step is to form the ceramic particles into a desired shape. This is accomplished by the addition of water and/or additives such as binders, followed by a shape forming process. Some of the most common forming methods for ceramics include extrusion, slip casting, pressing, tape casting and injection molding. After the particles are formed, these "green" ceramics undergo a heat-treatment (called firing or sintering) to produce a rigid, finished product. Some ceramic products such as electrical insulators, dinnerware and tile may then undergo a glazing process. Some ceramics for advanced applications may undergo a machining and/or polishing step in order to meet specific engineering design criteria.

The processing of glass products is different from ceramics. In glass production, raw materials such as silica, lime, and soda ash are melted in a furnace, then formed into the desired shape (i.e. pressed plate, fibers, molded bottle, plate glass, etc.) while still molten. After the molten glass is formed, it is quickly cooled, "freezing" the glass into place to form the finished product. The glass typically undergoes additional processing steps such as cutting, etching, coating, grinding, decorating, or heat treating (tempering).

Some of the challenges that face ceramic and glass manufacturers, and suppliers include availability and cost of raw materials, quality and cost of labor, changing markets, quality control, capital for expansion, import pressure, and environmental, health and safety standards.

3.1.4 Structure and properties of ceramics

The properties of ceramic materials, like all materials, are dictated by the types of atoms present, the types of bonding between the atoms, and the way the atoms are packed together. This is known as the atomic scale structure. Most ceramics are made up of two or more elements. This is called a compound. For example, alumina (Al_2O_3), is a compound made up of aluminum atoms and oxygen atoms.

The atoms in ceramic materials are held together by chemical bonds. The two most common chemical bonds for ceramic materials are covalent and ionic. For metals, the chemical bond is called the metallic bond. The bonding of atoms together is much stronger in covalent and ionic bonding than in metallic. That is why, generally speaking, metals are ductile and ceramics are brittle.

Another structure that plays an important factor in the final property of a material is called microstructure. The microstructure of a material is the structure that can be seen using a microscope, but seldom with the naked eye. For ceramics, the microstructure can be entirely glassy (glasses only); entirely crystalline; or a combination of crystalline and glassy. In the last case, the glassy phase usually surrounds small crystals, bonding them together.

The atomic structure primarily effects the chemical, physical, thermal, electrical, magnetic, and optical properties. The microstructure also can effect these properties but has its major effect on mechanical properties and on the rate of chemical reaction. Due to ceramic materials wide range of properties, they are used for a multitude of applications. In general, most ceramics are:

(a) hard
(b) wear-resistant
(c) brittle
(d) refractory
(e) thermal insulators
(f) electrical insulators
(g) nonmagnetic
(h) oxidation resistant
(i) prone to thermal shock
(j) chemically stable

Of course there are many exceptions to these generalizations. For example, borosilicate glasses (glasses that contain silica and boron as major ingredients) and certain glass ceramics (glasses that contain a crystalline phase) and NZP ceramics are very resistant to thermal shock and are used in applications such as ovenware, stove tops and kiln furniture respectively. Also, some ceramics are excellent electrical conductors and an entire commercial market is based on the fact that certain ceramics (ferrites) are magnetic.

3.1.5 History of ceramics

Archeologists have uncovered human-made ceramics that date back to at least 24,000 BC. These ceramics were found in Czechoslovakia and were in the form of animal and human figurines, slabs, and balls. These ceramics were made of animal fat and bone mixed with bone ash and a fine claylike material. After forming, the ceramics were fired at temperatures between 500 — 800 °C in domed and horseshoe shaped kilns partially dug into the ground with loess walls. While it is not clear what these ceramics were used for, it is not thought to have been a utilitarian one. The first use of functional pottery vessels is thought to be in 9,000 BC. These vessels were most likely used to hold and store grain and other foods.

It is thought that ancient glass manufacture is closely related to pottery making, which flourished in Upper Egypt about 8,000 BC. While firing pottery, the presence of calcium oxide (CaO) containing sand combined with soda and the overheating of the pottery kiln may have resulted in a colored glaze on the ceramic pot. Experts believe that it was not until 1,500 BC that glass was produced independently of ceramics and fashioned into separate items.

Since these ancient times, the technology and applications of ceramics (including glass) has steadily increased. We often take for granted the major role that ceramics have played in the progress of humankind. Below are just a few examples of how important ceramics are to society.

3.1.6 Impact on society

Refractories

We all know that metals and their alloys are used to make automobiles, machinery, planes, buildings, and a myriad of other useful things possible. But metal production would not be possible without the use of a ceramic material called refractories. Refractories can withstand volatile and high-temperature conditions encountered in the processing of metals. Refractory ceramics are enabling materials for other industries as well. The chemical, petroleum, energy conversion, glass and other ceramic industries all rely upon refractory materials.

Uses in the construction industry

The multi-billion dollar construction industry encompasses areas such as commercial buildings, residential houses, highways, bridges, and water and sewer systems. These areas, which we often take for granted, would not be possible without ceramic materials. Products such as floor, wall and roofing tile, cement, brick, gypsum, sewer pipe, and glass are the building blocks in the world of construction.

What would our homes and businesses be if we did not have windows? The applications of glass in the construction industry include various types of windows to let in natural light. Approximately three billion square feet of glass is produced each year to make various types of windows. To put things in perspective, imagine a 200-foot wide glass highway stretching from New York to Los Angeles. The different types of glass for windows include safety, stained, tinted, laminated and non-reflective. Additionally, glass fibers are used for insulation, ceiling panels and roofing shingles, helping us stay warm and dry.

Clay brick is used to build houses and commercial buildings because of its strength, durability, and beauty. Brick is the only building product that will not burn, melt, dent, peel, warp, rot, rust or be eaten by termites. Brick comes in approximately 10,000 different colors, textures, sizes, and shapes. Ceramic tile is used in applications such as flooring, walls, countertops, and fireplaces. Tile also is a very durable and hygienic construction product that adds beauty to any application.

Lighting electrical

An important invention that changed the lives of millions of people was the incandescent light bulb. This important invention by Thomas Edison in 1879 would not be possible without the use of glass. Glass properties of hardness, transparency, and its ability to withstand high temperatures and hold a vacuum at the same time made the light bulb a reality.

The evolution of lighting technology since this time has been characterized by the invention of increasingly brighter and more efficient light sources. By the middle of twentieth century, methods of lighting seemed well established — with filament and fluorescent lamps for interiors, neon lamps for exterior advertising and signs, and sodium discharge lamps for streets. Since this time, light-emitting diode (LED) technology has been developed with applications in watches, instrument panel indicators, telecommunications (optical fiber networks), data storage (CD technology), and document production (laser printers).

Electrical applications

The vast electronic industry would not exist without ceramics. Ceramics have a wide range of electrical properties including insulating, semi-conducting, superconducting, piezoelectric and magnetic. Ceramics are critical to products such as cell phones, computers, television, and other consumer electronic products.

Magnetic storage of data has developed in parallel with semiconductor computer chips and has been equally vital to computing and information handling. Without magnetic storage there would be no Internet, no personal computers, no large databases and no gigabytes, terabytes and exabytes of data which computers now manipulate. Today more than 150 million hard disc drives and 50 million videocassette recorders are made annually worldwide.

Ceramics are even being used to enhance our sporting activities. Piezoelectric ceramics (piezoceramics) are being used to make "smart" sporting goods equipment. That is, sporting goods that can respond to its surrounding environment in order to increase its effectiveness. Uses include snow skis, baseball/softball bats, and shock absorbers in mountain bicycles.

For example, K2 Corporation uses a control module produced by Active Control Experts inside a line of its skis. The control module contains a piezoceramic material that dampens vibrations from the ice and snow, helping keep the skis on the snow and thus enhancing stability, control, and ultimately speed. Worth, Inc. now has a line of bats that incorporates a similar module. The piezoceramic control module helps to increase the bats sweet spot and reduces unwanted sting on off-center hits.

Ceramic spark plugs, which are electrical insulators, have had a large impact on society. They were first invented in 1860 to ignite fuel for internal combustion engines and are still being used for this purpose today. Applications include automobiles, boat engines, lawnmowers, and the like. High voltage insulators make it possible to safely carry electricity to houses and businesses.

Communication

Glass optic fibers have provided a technological breakthrough in the area of telecommunications. Information that was once carried electrically through hundreds of copper wires can now be carried through one high-quality, transparent, silica (glass) fiber. Using this technology has increased the speed and volume of information that can be carried by orders of magnitude over that which is possible using copper cable.

Optical fibers are a reliable conduit for delivering an array of interactive services, using combinations of voice, data and video. Whether it's multimedia and video applications, high-speed data transmissions and Internet access, telecommuting or sophisticated, on-demand services, optical fibers make it easier to communicate.

Medical

Ceramics are becoming increasingly useful to the medical world. Surgeons are using bioceramic materials for repair and replacement of human hips, knees, and other body parts. Ceramics also are being used to replace diseased heart valves. When used in the human body as implants or even as coatings to metal replacements, ceramic materials can stimulate bone growth, promote tissue formation and provide protection from the immune system.

Dentists are using ceramics for tooth replacement implants and braces. Glass microspheres smaller than a human hair are being used to deliver large, localized amounts of radiation to diseased organs in the body. Ceramics are one of the few materials that are durable and stable enough to withstand the corrosive effect of bodily fluids.

Imaging systems are critical for medical diagnostics. Modern ceramic materials play an important role in both ultrasonic and X-ray computed tomography (CT) systems. Transducers utilizing lead zirconate titanate (PZT) based piezoelectric ceramics are the heart of ultrasonic systems. Transducers generate the ultrasonic acoustic waves and detect the reflected signals to form the image.

Ultrasound can be used to examine many parts of the body including the abdomen, breasts, female pelvis, prostate, thyroid and parathyroid, and the vascular system. Most commonly, ultrasound is used during the first, second, or third trimester of pregnancy. New developments in ultrasound now enable doctors to see reliable images of ulcerated plaque or irregularities in the blood vessels — something doctors have never been able to see before. This may allow doctors to better identify those at risk for stroke because of its ability to more clearly illustrate the flow of blood, the walls of the carotid artery, and the movement of plaque.

X-ray CT scans are now a common diagnostic procedure in hospitals and clinical sites to image selected regions inside the human body for detection of cancer and other diseases. For CT scans, an X-ray detector is a crucial component that must have high efficiency to obtain high quality images. In 1988 GE Medical Systems unveiled an ultrahigh-performance detector with a breakthrough ceramic scintillator which gives better images using lower X-ray doses to the patient.

Environmental and space applications

Ceramics play an important role in addressing various environmental needs. Ceramics help decrease pollution, capture toxic materials and encapsulate nuclear waste. Today's catalytic converters in trucks and cars are made of cellular ceramics and help convert noxious hydrocarbons and carbon monoxide gases into harmless carbon dioxide and water. Advanced ceramic components are starting to be used in diesel and automotive engines. Ceramics light-weight, high-temperature and wear resistant properties results in more efficient combustion and significant fuel savings. Ceramics also are used in oil spill containment booms that corral oil so it can be towed away from ships, harbors, or offshore oil drilling rigs before being burned off safely.

Reusable, lightweight, ceramic tile make NASA's space shuttle program possible. The 34,000 thermal barrier tiles protect the astronauts and the shuttles aluminum frame from the extreme temperatures (up to approximately 1,600 °C) encountered upon re-entry into the earth's atmosphere.

While the list could go on, one can begin to grasp the critical impact that ceramics have played in our past and present and will continue to play in the future.

3.2 What are advanced ceramics

Ceramics are traditionally described as inorganic, nonmetallic solids that are prepared from powdered materials, are fabricated into products through the application of heat, and display such characteristic properties as hardness, strength, low electrical conductivity, and brittleness. Ceramics can be divided into two categories (see Figure 3.1). Advanced ceramics represent an "advancement" over this traditional definition. Through the application of a modern materials science approach, new materials or new combinations of existing materials have been designed that exhibit surprising variations on the properties traditionally ascribed to ceramics. As a result, there are now ceramic products that are as tough and electrically conductive as some metals.

With the development of advanced ceramics, a more detailed, "advanced" definition of the material is required. Since 1993 an advanced ceramic is described as "an inorganic, nonmetallic (ceramic), basically crystalline material of rigorously controlled composition and manufactured with detailed regulation from highly refined and/or characterized raw materials giving precisely specified attributes." A number of distinguishing features of advanced ceramics are pointed out in this definition. First, they tend to lack a glassy component; i.e., they are "basically crystalline". Second, microstructures are usually highly engineered, meaning that grain sizes, grain shapes, porosity, and phase distributions (for instance, the arrangements of second phases such as whiskers and fibres) are carefully planned and controlled. Such planning and control require "detailed regulation" of composition and processing, with "clean-room" processing being the norm and pure synthetic compounds rather than naturally occurring raw materials being used as precursors in manufacturing. Finally, advanced ceramics tend to exhibit unique or superior functional attributes that can be "precisely specified" by careful processing and quality control. Examples include unique electrical properties such as superconductivity or superior mechanical properties such as enhanced toughness or high-temperature strength. Because of the attention to microstructural design and processing control, advanced ceramics often are high value-added products.

Advanced ceramics are also known as engineering ceramics, technical ceramics or high performance ceramics and referred to materials which exhibit superior mechanical properties, corrosion/oxidation resistance, and thermal, electrical, optical or magnetic properties. Advanced ceramic materials constitute an emerging technology with a very broad base of current and potential applications and an ever-growing list of material compositions. The applications can be categorized into structural ceramics, functional ceramics. Components which are subjected mainly to mechanical loads (tensile and pressure strain, bending moments etc.) fall under the category structural ceramics. Sometimes the terms industrial or engineering ceramics are also used for these products. Components which fulfill an electrical, magnetic, dielectrical, optical etc. function belong to group of functional ceramics. The followings are some applications of advanced ceramics:

Structural (Engineering) ceramics

(a) cutting and grinding tools
(b) engine components such as pistons and valves
(c) refractories and thermal coatings
(d) glass and glass-ceramics
(e) filters and membranes
(f) ball bearings and wear-resistant linings
(g) bioceramics, including implants and artificial joints

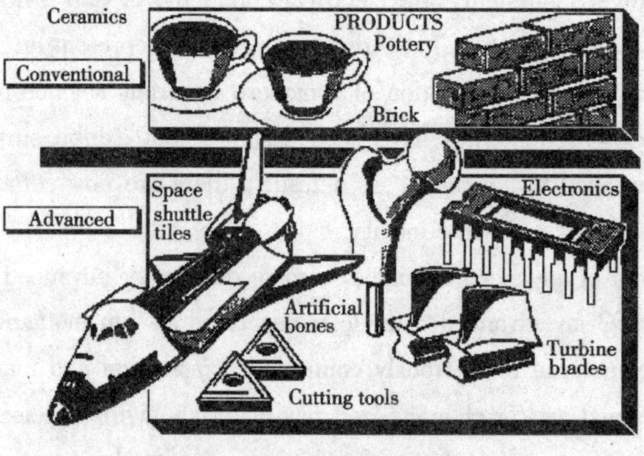

Figure 3.1 The categories of ceramics

Functional ceramics

(a) electronic components such as resistors, capacitors, varistors
(b) ionic and mixed ionic/electronic conductors for gas sensors, fuel cells
(c) solid-state batteries
(d) high temperature superconductors
(e) magnetic ceramics used in cassette tapes, computer memory, transformers, loudspeakers
(f) laser crystals used in CD and DVD players, medical devices
(g) piezoelectrics, used as actuators and transducers, ultrasonic generators
(h) pyroelectrics and electro-optics used in night-vision goggles, motion and infra-red detectors

Advanced ceramics encompass a variety of materials. The materials include: oxides, carbides, nitrides, borides, silicates and glass ceramics. The most commonly used advanced ceramic materials are: alumina, zirconia, silicon carbide, silicon nitride, sialon, ferrites and titanates.

3.3 Structural ceramics and its development

3.3.1 Introduction and current status

Scope of the subject

Structural ceramics are taken to include oxides (Al_2O_3, ZrO_2, MgO, SiO_2, TiO_2), nitrides (Si_3N_4, sialons, AlN, BN, TiN), carbides (B_4C, SiC), borides (cubic BN, TiB_2), and composites based on the above (CMCs) involving reinforcement with particulate or fibrous ceramic phases to achieve either enhanced toughness, or improved thermal or electrical conductivity.

Current applications include automotive components (tappet inserts, precombustion chambers, glow plugs, turbo-rotors, brakes), wear parts, bearings, metallurgical and molten metal handling tools, cutting inserts and grinding materials, pump and valve components, textile thread guides, body armour and other protective security devices, furnace ware, laser bodies, thermal spraying refractories, high-performance insulators such as substrates and microwave devices, knives and shears, cores and moulds for investment casting, liquid and dust filters, catalyst supports, zeolite products, bioengineering materials, e.g. femoral and knee prostheses.

Industrial developments

For a number of materials types, the technology could be considered mature by the 1970's, but during the 1970's and 1980's the UK engineering ceramics industry experienced a significant "technological push". In parallel with heavy investment in the USA, Japan and Germany, several speculative investments were made, including initial work on CMC's. In part this was supported by DTI programmes ACT (advanced ceramics in turbines) and CARE (ceramics applied to reciprocating engines) which promoted material development and collaboration between all sectors. Since this period some newer products have been commercialised for niche markets, e.g. zirconias and silicon nitrides, while markets for some of the more traditional applications, e.g. high-power insulators have dwindled. The industry has gone through a period of rationalization, consolidation and internationalization such that at the beginning of the 21st century it is polarized between a few medium sized companies, which are parts of large international companies (only one of which is UK based) and several SME's. Much of these companies markets are in Europe.

3.3.2 Trends

Research/Development

The UK boasts some nine or ten university groups with "internationally competitive" research programmes on structural ceramics (including CMCs), the programmes being funded either by EPSRC, EC programmes or by UK industry. Much of the earlier research and development was stimulated by predictions of extensive ceramic substitution of metallic components in aerospace and automotive applications. The silicon nitrides and their derivative "sialons", both of which were based on internationally leading research in the UK, have been especially prominent in the development of engineering ceramics. R & D has followed international trends, but in the UK has not been supported by the large programme funding typified by USA and Japanese initiatives of the 1970/80s. The UK groups have contributed significantly to the science underlying the control of microstructure in both nitride and oxide (mainly zirconia-based) engineering ceramics and there is now a substantial knowledge-base in the linkage of mechanisms for deformation and fracture with microstructure.

A persistent problem, which has inhibited technology-transfer in applications involving "critical" engineering components, is the toughness (K_c) limitation for monolithic ceramic microstructures. This has stimulated a series of research programmes on ceramic composites (CMCs), initially in the form of particle and whisker dispersions within ceramic matrices and later by the introduction of semi-continuous ceramics fibers.

The use of dispersed phases to enhance K_c has not been very successful; values rarely exceed 10 $MPa \cdot m^{1/2}$ except for additions of ZrO_2 to oxide matrices in which transformation-toughening mechanisms operate. Additions of SiC whiskers to $A1_2O_3$ did provide for critical K_c increments from ~4 to 9 which facilitated the commercial development of cutting inserts. Modelling of toughening mechanisms in this system (largely crack bridging/pull-out) provided useful feedback to a significant development in monolithic Si_3N_4 based ceramics; that of "self-reinforced" microstructures with large prismatic (Si_3N_4) crystals. Dispersions of thermally and electrically conducting phases (TiN, TiB_2, etc.) have also provided useful secondary properties of enhanced thermal shock or electro-discharge-machinability (EDM). A negative feature of "dispersion toughening" is a reduction in ease of processing and fabrication, normally requiring pressure-assisted sintering (HP or HIP).

A recent development, pioneered by Niihara and co-workers in Japan, has been the emergence of "ceramic nanocomposites", defined as composites in which at least one component is submicron in scale. The improvements claimed for most such microstructures have been in properties such as strength, creep resistance and wear resistance, rather than in K_c. Despite conflicting and irreproducible early results, it is now clear that some nanocomposites do show useful, novel, and reproducible property improvements compared with their monolithic counterparts.

The most-studied system has been alumina (with grain size typically ~3 μm containing a dispersion of submicron (typically 200 nm) SiC particles). Alumina/SiC nanocomposites possess a much better surface finish after grinding, and are more resistant to "severe" wear than pure alumina. The underlying mechanisms have not been uniquely identified, but have their origins in the ability of the SiC "nanoparticles" to impede the nucleation and propagation of cracks on alumina grain boundaries. Alumina/SiC nanocomposites are approaching industrial exploitation and progress in the UK is now being made in developing commercially viable production routes such as pressureless sintering. Alternative compositions have received less attention.

The problem of toughness together with the development of continuous ceramic fibres from polymer-spun precursors has stimulated the more recent evolution of fibrous CMCs. These are, ideally, damage-tolerant microstructures in which load is transferred to semi-continuous fibres during matrix microcracking (or creep at high temperatures). The most significant developments internationally, have been in a range of SiC fibres and in CVI and DIMOX (Al_2O_3) matrix infiltration from which a limited range of products have developed. Major problems, which persist, have been cost and the inability to produce low-cohesion interphases which are oxidation-resistant (carbon and boron nitride have been most frequently used). Recent (oxide fibre/oxide interphase + matrix) systems currently lack adequate high temperature thermo-mechanical stability for intended applications and are difficult to fabricate from woven-fibre precursors.

Industry

Over the last three decades, the structural ceramics industry has taken a number of the more cost-effective material developments through to the market place. In order to do this, it has been reliant on the commercial supply of consistent powders stemming from research programmes, and therefore it could be perceived to be slow to capitalise on research developments, e.g. in nanocomposites. However, this would be an unfair perception bearing in mind the cost pressures experienced by the industry when attempting to penetrate markets previously dominated by other materials. Nevertheless, viewed overall, the diversity of materials now available and the fields in which they are used is remarkable.

The U.K. finds increasing difficulty in competing internationally, because most industries are too small to take advantage of the current international trend of establishing R&D and production facilities in the most cost-effective areas world-wide, whilst distributing sales outlets in major customer locations. Industry has also been negatively influenced by the lack of appropriately skilled materials engineers and technicians, which is another reason why companies are also looking abroad as a way of further developing their operations. UK structural ceramic companies are experiencing growth, but at a slower rate than overseas competitors (especially Japan, USA).

A major obstacle to further market penetration for engineering ceramics continues to be that of component cost, based on the need for high purity nanosize powder raw materials, complex processing and high temperature fabrication. This is especially relevant to the large automotive sector, exemplified by the proven technology (largely in Germany) of engine valves that are currently produced on a pilot scale at a cost approximately twice that considered economically viable.

The industry has made efforts to improve its marketing of new products, its quality assurance, and the reliability of product data. To improve the pedigree of structural ceramic products, the industry has strongly supported programmes to standardize terminology, classification and testing methods through BSI, CEN and ISO, actions that are beginning to pay dividends in terms of customer confidence. However, much remains to be done in the education of the engineering community in the successful mechanical and thermomechanical use of brittle materials.

3.3.3 Future work strategies

Markets

(a) Focus on demonstrator components with quantified property targets which are not too ambitious (and use existing ceramic material types where possible).

(b) Strengthening of links with engineers, with a view to applications in areas away from aerospace (automotive, metal/glass processing, building, health care, environmental).

General engineering

(a) Improve the status of engineering ceramics as engineering materials by improved provision of property and performance data through international standards procedures, and improved understanding of design procedures for relatively brittle materials.

(b) Development of improved, cheaper and more tailored raw materials. Development of innovative processing techniques for both monolithics and CMCs (e.g. for CMCs, "fibrous-monolith" and "freeform processing" projects currently on-going in the USA).

(c) Development of better dimensional control in powder processing to achieve required shape with minimal iterations in order to improve competitiveness.

Materials

(a) Strengthening of links with chemists with respect to colloidal/sol gel syntheses and handleability of nanoscale powders for sinterability and ultra-fine grain-sized products. Critically assess the potential (including production costs) of hydrothermal syntheses of ceramic powders, based on significant initiatives in Japan (e.g. creation of a Research Institute for Solvothermal Technology in Takamatsu). N.B. These are generic needs covering functional and bioceramics.

(b) Development of high temperature fibres and CMC interphases, for aerospace and power generation, with long term stability up 1,350 ℃ in oxidizing conditions.

(c) Development of thermal protection systems to operate CMCs up to ~1,600 ℃.

(d) Development of joining technologies (ceramic-metal and ceramic-ceramic).

(e) Modelling of sintering/microstructure development.

3.4 Electroceramics and its developing directions

3.4.1 What are electroceramics

The term electroceramic is used to describe ceramic materials that have been specially formulated for specific electrical, magnetic or optical properties. Their properties can be tailored to operation as insulators, ferroelectric materials, highly conductive ceramics, electrodes as well as sensors and actuators.

To clarify, ceramic materials are mainly polycrystalline, non-metallic inorganic substances typically prepared by solid state chemical reaction followed by a sintering process at elevated temperatures. The key features that generally describe ceramics, and consequently electroceramics are:

(a) The presence of ionic-covalent bonding.

(b) Microstructures comprising inorganic crystal compounds and/or amorphous glass in varying proportions.

(c) Thermal processing conducted at elevated temperatures.

3.4.2 Common applications for electroceramics

(a) high dielectric constant capacitors

(b) piezoelectric sonar transducers

(c) ultrasonic transducers

(d) radio & communication filters

(e) pyroelectric security surveillance devices

(f) medical diagnostic transducers
(g) stereo tweeters, buzzers, gas ignitors
(h) PTC sensors and switches
(i) ultrasonic motors
(j) electro-optic light valves
(k) thin-film capacitors
(l) ferroelectric thin-film memories

3.4.3 Fundamental research needs and developing trends for electroceramics

Electroceramics are responsible for scientific and technological innovations which have contributed immeasurably to modern society. The zirconia oxygen sensor and 3-way automotive catalyst have enabled vast improvements in air quality; piezoelectric ceramic actuators have enabled technologies ranging from advanced sonar to scanning probe microscopies to the Hubble Telescope repair; ceramics packaging and multilayer device technology have facilitated the continuing miniaturization of electronics, and microwave dielectrics enable present-day wireless communications.

These applications are based on electrical, electromechanical, dielectric, electrochemical and chemical functions. The science involved is inherently multidisciplinary. What has distinguished the contribution of ceramics is the recognition that the performance of real materials is determined not only by intrinsic structure and composition, but also by defects, interfaces, and often-complex microstructures, each influenced by processing.

The working group discussed a number of areas in which historically close ties between electronic ceramics and device technologies exist. A specific technology focus is warranted for some areas (e.g., wireless communications) due to the size of the field and the potential impact. In other areas, the research opportunities cut across many fields of application. As discussed further in the section role of NSF in supporting collaborations with industry and other sectors, teaming between university researchers and industry is particularly recommended where ceramics for electronics are concerned, in order to facilitate the informed selection of basic research projects for technological impact.

It was also found that in several major technologies, ceramics are enabling materials, yet have not historically enjoyed the widespread participation of the ceramics community. Examples include heterogeneous catalysis and photocatalysis, in which the active materials are frequently semiconducting oxides, and lithium ion battery technology, in which oxides are the active cathode materials. These represent areas where ceramics can and should have a major impact in the future.

The following topics were discussed in greater detail as examples where fundamental research is particularly needed.

Size effects in electroceramics

Electroceramics are used as thin films, single crystals and bulk polycrystals. In device technologies, trends towards miniaturization and integration with silicon have resulted in an increasing focus on the properties and processing of thin films. The properties of thin film ceramics, often oxides, can deviate remarkably from those of their bulk counterparts. The understanding of such properties is an important need and opportunity. A specific example discussed during the workshop concerns high permittivity ceramics used in DRAM's, where it is found that in very thin films, dielectric properties can be dominated by interfacial rather than bulk capacitance. Predictors of performance based on bulk properties are consequently of limited validity.

Chemical and electrochemical applications also use thin film (e.g., fuel cells, sensors) and fine powder ceramics (e.g., catalysts). Size effects can be equally important in these applications. Research in this area is therefore generally concerned with understanding properties under severe size and shape constraints. Properties and phenomena of interest include:

(a) Dielectric response of thin films and interfaces

(b) Nonstoichiometry and point defects in thin films and small crystallites

(c) Phase stability and phase transitions

(d) Electronic and ionic transport

(e) Reliability; mechanisms of device degradation

(f) Electromechanical response of thin films

It is anticipated that improved fundamental understanding in this area will have a fairly direct impact on microelectronics and communications technologies, in which ceramics are used in conductive, magnetic and optical applications.

Ceramics for wireless communications

Communication technologies and in particular, wireless communication technologies, are one of the most important growth businesses in the global electronics industry. Ceramic requirements; specifically for filters, oscillators and resonators; are critical and represent the single largest cost and size driver of the end devices. Most new ceramics enabling significant advances are coming from research done overseas. Without an increase in effort, the US risks losing a leadership position in both the research and technology.

Ceramics contributes to this technology in several ways. There are several classes of microwave ceramic dielectrics which offer the combination of relatively high permittivity (ε_r = 20 — 100), low loss (Q = 1,000 — 20,000) and low temperature coefficient of frequency ($\tau_f \cong 0$), all at microwave frequencies (1 — 4 GHz). These materials have permitted size reductions in cell site transmitters and receivers, hand set transceivers and satellite communication filters. Also, high temperature ceramic superconductors are finding their first large scale commercial uses in very low loss microwave resonators for wireless communications. These very low loss resonators allow the manufacture of multipole filters with very high frequency selectivity and very low insertion loss for application in environments where nearby frequency interference causes performance problems.

Research opportunities abound in this field. Several top materials R & D issues have already been identified by U. S. Industries. These are, in prioritized order:

(a) Develop a fundamental understanding of the relationships between chemistry, structure (including texture) and dielectric properties of microwave ceramic dielectrics.

(b) Development of phase diagrams of pertinent ceramic systems which are accurate and reliable.

(c) Research directed toward the discovery of new ceramic compositions with $\varepsilon_r > 150$, $\tau_f \cong 0$ and high Q at microwave frequencies.

(d) Understanding of the reaction kinetics during reactive sintering.

Beyond this, the Working Group identifies the following basic issues:

(a) Development of a basic understanding of the physical basis to stretched exponential decay times responsible for higher polarizabilities than those expected from ionic contributions.

(b) A "microscopic" understanding of the dielectric constant and conductivity (loss factor, Q) at high frequencies. Recent new models have been proposed for the losses in glasses in terms of multi-atom (jelly fish) movements. Similar models may apply to crystalline ceramics, and should at least be tested as a starting point for understanding the underlying phenomena.

(c) Understanding the role of composition on Q, ε_r and τ_f. This includes an understanding of the effect of dopants on polarizability and its coupling to high frequencies.

Integrated functions in electronic ceramics

The electronic ceramic industry has derived tremendous value from the integration of metals and insulators (multilayer ceramic packages); metals and high e ceramics (multilayer ceramic capacitors); and metals and piezoelectric or electrostrictive ceramics (actuators). Other multifunctional electronic ceramic devices and systems can benefit from the economies of manufacturing cost and size reduction provided by integration. A few actively pursued examples include:

(a) high ε dielectrics, metals and magnetic inductors (LC circuits)

(b) high ε dielectrics, metals and cofired resistors (RC circuits)

(c) insulators, high μ magnetics and metals (small surface-mount inductors and transformers).

The technological potential in this field is very high. Examples of new devices include combined sensors and actuators, power supplies integrated on silicon chips, and biomedical systems. Ceramics as a class of materials is critical to most of these electronic functions since it offers the broad array of properties needed, including high magnetic permeability, high dielectric permittivity, high or low dielectric constant, tailorable resistivities (including superconductivity), hermaticity, low dielectric or magnetic loss, electrooptic functionality, high piezoelectric or magnetrostrictive strain, and outstanding magnetoresistive properties.

The greatest opportunities in this area lie in understanding how to process dissimilar materials while achieving the desired integration of functions. Co-fired devices are an example where the ceramics field has made the seminal contributions in the past. Current needs in this area include:

(a) Identifying systems of dissimilar dielectric magnetic, resistive and piezoelectric ceramics which can be co-sintered in the 800 to 1,000 ℃ range with acceptable properties. This includes resolving issues of interdiffusion and thermomechanical properties mismatch as well as controlling sintering shrinkage as a function of time and temperature.

(b) Developing forming methods to accomodate disparate shrinkage during firing. The advantages provided by the ductility of metals will not be present in co-sintering of many multifunctional electronic ceramics.

Chemically and electrochemically active ceramics

Chemically active ceramics are widely used in chemicals production, sensing, and energy production, conversion, and storage. Specific examples include:

(a) Solid oxide fuel cells for energy production.

(b) Intercalation compounds in rechargeable Li-ion batteries.

(c) Ceramic gas sensors for monitoring O_2, CO, NO_x, and hydrocarbon emissions.

(d) Titania for the photocatalytic remediation of waste water.

(e) Mixed metal oxides for the synthesis of chemical intermediates.

(f) Perovskites for gas separation membranes (e.g., $(La_{1-x}Sr_x)(Fe_{1-y}Co_y)O_3$).

(g) Zeolites and clays for petroleum refining.

(h) Alumina catalyst supports and ceria oxygen buffers in the 3-way automotive catalyst.

Core topics in ceramics basic research such as defects, electronic and ionic transport, crystallography, microstructure, surfaces, and interfaces all have impact on this class of materials. In some areas such as heterogeneous catalysis, photocatalysis, and lithium battery technology, the field of ceramics has been surprisingly underrepresented, given that the active materials are ceramics. These are areas in which the materials science approach can have important impact.

The Working Group found that many of the most exciting opportunities are concerned with surface active materials. Broadly speaking, the key goal of researchers in this area is understanding how to control the chemical activity and selectivity of ceramic surfaces and interfaces by manipulating their structure and composition. A few specific topics include:

(a) Understanding how second phases and hetero-interfaces influence the selectivity of sensors. For example, why does the addition of Y_2O_3 to TiO_2 (anatase) make it selective for CO while the addition of Al_2O_3 makes it selective for H_2?

(b) Understanding the mechanisms controlling catalytic anisotropies and using this phenomenon to influence chemical functionality. For example, why is the (100) facet of vanadyl pyrophosphate active for the oxidation of *n*-butane to maleic anhydride while the others are not?

(c) Stabilizing high surface area materials against phase transformations and consolidation. For example, how are γ-alumina supports stabilized against transformation to α-alumina in automotive catalysts?

(d) Understanding how surface electrochemical potential influences catalytic activity and selectivity. Examples include the enhanced catalytic oxidation of hydrocarbons and CO by using an electrically biased solid electrolyte such as Y_2O_3-ZrO_2.

(e) Improving the efficiencies of photcatalytic reactions on metal oxide surfaces. For example, what intrinsic or defect properties of TiO_2 influence the efficiency with which photo-generated charge carriers are transferred from the surface to adsorbed reactants?

Another particularly timely area is the efficient design of new materials, by computational and/or experimental approaches. This includes the discovery of new compounds as well as the design of composites and integrated systems. Examples include:

(a) Design of improved intercalation compounds for lithium ion batteries and electrochromic devices. For instance, what are the microscopic mechanisms of cycling fade in Li-ion batteries, and can they be circumvented through crystal chemical or microstructural engineering of the active materials?

(b) Design of composites for electrochemical systems. For example, can the composite which best satisfies the multiple demands placed on fuel cell components of thermomechanical compatibility, thermodynamic stability, and gas-phase and solid-state transport, be designed from basic principles rather than by empiricism?

Chemically active ceramics is, like electronic ceramics, a highly interdisciplinary field, being studied not only by ceramists but also by electrochemists, solid state chemist, catalytic chemists, chemical engineers, and surface scientists. The uniqueness of the ceramics field has been its emphasis on processing-structure-property relationships, which is not present in (for example) surface physics or crystal chemistry. For greatest impact, the ceramics community should attempt to strengthen its interactions (both through interdisciplinary research, professional societies, and funding agencies) with these allied fields.

3.5 Bioceramics

3.5.1 Introduction

Ceramics encompass such a vast array of materials that a concise definition is almost impossible. However, one workable definition of ceramics is: an inorganic and nonmetallic material. Ceramics can be divided into two classes: traditional and advanced. Traditional ceramics include clay products, silicate glass and cement; while advanced ceramics consist of carbides (SiC), pure oxides (Al_2O_3), nitrides (Si_3N_4), non-silicate glasses and many others. Ceramics offer many advantages compared to other materials. They are harder and stiffer than steel; more heat and corrosion resistant than metals or polymers; less dense than most metals and their alloys; and their raw materials are both plentiful and inexpensive.

The use of prosthetic implants has been increasing rapidly since their conception near the turn of the century due to a rapid increase in our understanding of biological processes. Early studies showed that several materials were accepted by biological systems by two different mechanisms, the immune system, and the toxicity of the implanted substance to the rest of the internal environment. A material is said to be bioinert if it produces little or no physiological reaction from its host. These reactions are usually the immune system rejecting the material. Gold is a classic example. It has been used as a bioinert material for the last 4,000 years in items such as fillings and "replacement" body parts. A material is bioactive if a physiological reaction occurs which incorporates the material into the local system. Other foreign materials that either poison the system or produce an immune reaction are often found as impurities in ceramics.

There are several polymeric and metallic biocompatible materials with which bioceramics are often matched. Some will argue that the biocompatability of metals depends on their passivation with a thin oxide layer, which is by definition a ceramic. It is known, however, that the concentration of metallic ions in the body is proportional to overall surface area of an implant. Even the widely applied titanium-based alloy with some aluminum and vanadium (Ti-6Al-4V) has been shown to release some of each of metallic species to blood and tissue. Surprisingly, the human body develops some adaptation to the invaders; this response is related to the rate of release of corrosion products, as well as their toxicity. In vivo (in an active biological system) tests have shown that after a certain time period (depending on the host and many other variables) the corrosion products accumulate around the metallic implant and act as a buffer, slowing the release of any more corrosive products. In theory, after a long enough period an equilibrium would be reached and the release of the various ions would slow to almost nothing. The components of several metallic biomaterials (iron, nickel, chromium, cobalt, vanadium) have been indicated as causing an undesirable response in the human body that is similar to heavy metal poisoning.

Bioceramics are a growing industry and will continue to support a great deal of fundamental research. Due to the generous funding available to medical researchers, developments in the processing of ceramics are often made on biomaterials/bioceramics. These developments then are incorporated into other industries. The difficult biological environment has led to novel testing techniques and improvements in ceramic cleanliness. Alumina, zirconia and calcium phosphates have been found to be very useful when coupled with stronger but less biocompatible materials. This has led to the development of a wide variety of medical prostheses and has improved the lives of many people.

3.5.2 Bioinert and bioactive materials

A prerequisite for any synthetic material to be implanted in a body is its biocompatability, i.e. no inflammatory, or generally adverse reaction from the body. Additionally, an implant is expected to withstand loads (chemical, thermal and mechanical) without substantial deterioration or catastrophic event (reaction, fracture, abrasion) or alteration to the environment. The materials must not interact with blood, (e.g. produce clotting or denaturing of plasma proteins.)

Bioinert materials do not cause any response from their host. This is usually because they do not chemically react with the local chemicals due to the fact that the bioinert materials have already reacted with another (more reactive) substance. This is why most bioinert materials are oxides of metals and metalloids. As a result cells can live adjacent to the material but do not form a bond with it. Often fibrous protective cells grow near the implant surface to protect local cells from mechanical damage. Bioinert materials were the first used for prosthetics. Bioinert ceramics include alumina (Al_2O_3), zirconia ($ZrSiO_4$) and titanium oxide(TiO_4). These materials can be very strong but have the disadvantage of not bonding to the local cells. Numerous problems have been encountered in anchoring the bioinert implants to bone. In early implants, some implants became deformed or displaced, causing serious damage to the surrounding tissue.

Bioactive materials have been a more recent development. They are materials which can react in a positive way with the local cells to form bonds. Thus, they can be used to transfer loads to and from living cells. Normally, cells grow directly adjacent to the material and, in some cases, may actually grow into the material over time. Little is known about the mechanisms of bonding and resorption although Hench (1990) has shown that proteins and sugar can bond to bioceramics in a stable organized fashion over a range of chemical environments. Some interesting conjectures about how some bioceramics may have acted as catalysts to the formation of DNA like chains and thus the origin of life were made by Hench too. Howden (1980) has shown that blood vessels actually penetrate some bioactive prosthetics and that bone material can replace the prosthesis over time.

The primary bioactive ceramics are low silica glasses, and various calcium phosphates. These calcium phosphates are very similar to bone material. Bioactive ceramics are often made to be highly porous to increase their active surface. Uniform porosity of the order of 0.3 — 0.4 mm is desirable to facilitate the ingrowth of bone into pores in alumina (bioinert material) or hydroxyapatite (bioactive material).

The low strength and brittleness of many bioactive ceramics is their main disadvantage. Consequently they are most often used in conjunction with other stronger materials. Usually a bioactive ceramic is used to coat a bioinert metal, although many different combinations and morphologies has been used successfully. Fiber and particulate reinforcement are commonly used with bioactive materials as well.

Types of bioceramics are shown in Table 3.2.

Table 3.2 Types of Bioceramics — Tissue Attachment and Bioceramic Classification

Type of bioceramic	Type of Attachment	Example
1	Dense, nonporous, nearly inert ceramics attach by bone, growth into surface irregularities by cementing the device into the tissues, or by press fitting into a defect (termed morphological fixation).	Al_2O_3 (single crystal and polycrystalline)
2	For porous inert implants bone ingrowth occurs, which mechanically attaches the bone to the material (termed biological fixation).	Al_2O_3 (porous polycrystalline) Hydroxyapatite-coated porous metals
3	Dense, nonporous, surface reactive ceramics, glasses, and glass ceramics attach directly by chemical bonding with the bone (termed bioactive fixation).	Bioactive glasses Bioactive glass-ceramics Hydroxyapatite Bio-Glass Ceravital
4	Dense, nonporous (or porous), resorbable to be slowly replaced by bone.	Calcium sulfate (plaster of Paris) Tricalcium phosphate Calcium phosphate salts Calcium hydroxyapatite

3.5.3 Compatibility between bioceramics and the human environment

The major problem in designing implants is the fairly limited choice of materials, and consequently, determination of the compatibility of the material of choice with the tissue. There is no standard methods of compatibility testing and the number of variables involved is usually much larger than the typical engineering problem. For example, human blood is 1/3 as salty as sea water, stays at a steady 37 ℃, and contains active enzymes (the immune system). Consequently the human body is one of the most corrosive environments that inorganic substances can encounter. Furthermore, as the various metabolic processes occur in an organism the various complex molecules that may surround a substance perpetually change in concentration and variety. Lactic acid produced from muscle cells during anaerobic cellular respiration is a prime example. Additionally, the time factor of the "compatibility reaction" is important; the implant — tissue interaction is a sequential chain of reactions, characteristic for the material and the patient. Effectively, broad range of bioreactivity vs time chemical reactions result for various biomaterials. Calcium hydroxyapatite has been found to lose a large amount of its strength in the 20 to 30 month post implantation period.

Beyond the chemical factors, the response of tissue depends additionally on geometric characteristics of the implant, e.g. shape, size, surface/volume ratio. These factors will generally determine the state of stress at the interface and thus could interfere with the interfacial reactions. Porosity and its size distribution within an implant has been shown to affect the interactions. It has been generally established, that tissue will grow into pores larger than ~ 120 nm.

Most polymers seem to be slowly "digested" by the human body and metal are slowly corroded: high concentration of metallic elements has been detected close to the (metallic) implant surface. Passive oxide film can significantly slow down the reactions with tissue, rendering titanium or stainless steel virtually neutral. However, polymeric and metallic implants are generally classified as "temporary" implants, with very low adhesive strength of attachment to the tissue. Titanium is considered bioinert due to a thin layer of titanium oxide on its surface, although some studies show that titanium can be absorbed by the body in much the same way that mercury or other similar heavy metals are absorbed into the human body. Various researchers have found alarmingly high levels of titanium in test rats' nervous systems after the 20 week postoperative period.

On the other hand, bioactive materials, like Ceravital, bioglasses or hydroxyapatites, develop active intergrowth layer with adhesive strength of ~ 5 — 10 MPa. The bioinert ceramics, like ZrO_2, Al_2O_3, SiC, Si_3N_4 do not develop strong interfaces, but also do not release ions into the internal environment. This is at the expense of poorer mechanical performance and reliability of ceramics, as compared to metals. A compromise is ceramic coated metal, although some additional liabilities are created (e.g. adhesion of the coating; increased processing costs etc.).

The vitreous carbon (glass-like carbon) and carbon/carbon fibre composites are used for dental implants and heart valves because of their bio-compatibility, stability and simple manufacturing techniques. Vitreous carbon is produced through thermal degradation of organic polymers, such as phenol-formaldehyde (interestingly enough this was also the first fully synthesized plastic) usually as thin sections ~ 2 — 4 mm, or as coatings on metals. It is chemically inert, low density, hard and strong material. The elasticity of carbon/carbon composites in contact with bone has been indicated as major factor in c/c use for hip prosthesis. There are some suggestions that carbon implants can be carcinogenic if made improperly, e.g. some remnant aromatic groups are left over after production.

3.5.4 Most common bioceramics

Alumina (Al_2O_3)

Along with silica, alumina is the most widely utilised ceramic. Much of the research on this ceramic was done during the 1950s and 1960s. At that time it was investigated for its high strength and ability to be machined into almost any shape. It is produced from bauxites, impure aluminum hydroxides. Alumina is available in powder form in a very wide range of sizes, from fine (sub-micron) sinterable powders, to mm size grinding grits, and in fibrous and single-crystal form. It is a component of many enamels, glazes and glasses. Alumina is chemically stable against most environments except hydrofluoric acid and some molten salts. Alumina sets upon hydration if produced in the special form of re-hydratable alumina cement (more commonly in the form of calcium aluminate cement). Alumina is widely used for wear parts, abrasives, insulators, and medical implants.

High purity alumina (99.8% purity) that has been sintered at a minium temperature of 1,600 ℃ has a appearance much that of mother of pearl and is extremely smooth. This high purity alumina is used in partial and total hip replacement surgery, even without a lubricant other than water there is very little friction between alumina-alumina surfaces.

Hydroxyapatite ($Ca(PO_4)_3OH$ / $Ca_5(PO_4)_6OH_2$)

Hydroxyapatite is a complex calcium phosphate mineral, the chief structural element of vertebrate bone. Hydroxyapatite (HA) is also used in conjunction with $CaCO_4$ in the production of bone china. It is an important bioactive bioceramic. Many bioceramics (including glass, glass-ceramics and ceramics) develop a thin hydroxycarbonate-apatite (HCA) surface layer upon exposure to body fluids. Collagen fibrils incorporated within this growing inorganic mineral layer produce mechanically strong bond between the implant and the host tissue, to result in bioactive fixation. Pure HA does not have the physical characteristic of alumina. It is much weak and more susceptible to fracture but can be used as an additive to most other bioceramics.

Bioglass Mixtures (SiO_2, Na_2O, CaO, P_2O_5)

Since the discovery of stable SiO_2-Na_2O-CaO ceramic formulations several thousand years ago, most silicate ceramics used by man have an SiO_2 content of 65% (wt) or more. These SiO_2-Na_2O-CaO ceramics are more commonly called glass. Although these 65% silica glasses are extremely bioinert they contain few other desirable qualities. They are weak and shatter easily. The first bioactive glass contained 20% — 25% (wt) Na_2O and 20% — 25% (wt) CaO and only 45% — 55% (wt) SiO_2. Unfortunately this new type of glass was still very weak and brittle. With the addition of P_2O_5 to the SiO_2-Na_2O-CaO matrix this makes the glass extremely bioactive. Some of the patented ceramics such as "Ceravital" and "Bioglass 45S5" are so bioactive that within 1 hour of implantation a HCA layer (precursor to bone [in]growth) nearly 500 nm deep is formed. It has also been claimed that soft-tissue growth on the surface of "Bioglass 45S5" has been observed.

3.5.5 Uses for bioceramics

Specific bioceramics are best discussed in reference to specific applications. Many uses for bioceramics have been suggested but they are most accepted in dentistry and bone repair. Several examples are discussed below along with suggested materials.

Hip replacements

A large number of hip joints are replaced each year (over 200,000 worldwide since 1985). The joints require high strength components with low wear rates. They are made from bioinert materials although bioactive-bioinert composites have received recent attention. Long term studies by Ducheyne, *et al*. (1990) of up to ten years show success rates of about 90% for ceramic hips. Christel, *et*. *al*. (1988) analysed 17 retrieved alumina-alumina prostheses manufactured since 1977 (that had been implanted for periods from 17 — 77 months) and found notable particle tracks along both surfaces of the ball and joint, suggesting that a fair amount of friction and consequently wear occurred between ball and socket. It is logical to assume that more than water is need for a lubricant between the two surfaces.

Hip joints are a ball and cup type joint. The ball is usually made from alumina, titanium or more recently zirconia. The cup is usually made from a polymer although studies by Elssner, *et. al*. (1980) have shown low friction and wear rates with alumina cups. The parts are either epoxied, mechanically screwed or more recently cemented with bioactive ceramics to the bones. The hip bones are expected to grow into the pores to produce a strong bond. The prosthesis is initially attached with tricalcium phosphate paste which promotes the bone growth.

Dental inserts

The use of ceramics in dentistry is a multi-billion dollar industry in North America, involving many different proprietary materials. Two main uses are reported. Dental pins are devices which are drilled into the mandible and are used to hold prosthetic teeth in place. They are usually made from bioactive materials such as low silica glass or HA. They are used because of their ability to bond to biological tissues permanently. Due to their small size and ceramic material they are inherently weak and thus a wide range of composites have been experimented with, as reported by Soltesz (1988), such as carbon fiber reinforcement, ceramic coated steel pins and polymer toughened ceramics.

A large number of dental fillers are also used. A dental tooth filler must bond strongly to its support pin or tooth; it must have a high wear resistance; it must be tough; and it should match the natural tooth color closely. These considerations have led to the development of fuse cast alumina, machined single crystal, sintered zirconia and alumina inserts. These materials are then bonded with epoxy onto their support. Composites such as ceramic filled polymers and short fiber reinforcements have also been used.

Bone repair

When a bone suffers a trauma such as physical damage or surgery, a temporary mechanical support is often desired to allow the use of the bone while natural healing occurs. It is in this area that biodegradable bioactive materials such as tricalcium phosphates (TCP) are used. This ceramic is very low in strength and is often augmented with a stronger bioinert material such as titanium. The damaged area is repaired the same way a body shop fixes a panel with body filler. A structural support is coated with TCP and then glued into place with a bioactive cement or a bioinert epoxy. Bioinert screws are also used. Over time the bone will grow into the region once occupied by the TCP and take over the transfer of load to the structural support. Blood vessels actually grow into the ceramic prior to bone cell growth.

The basis of the bone bonding property of bioactive glasses is the chemical reactivity of the glass in body fluids. The surface chemical reactions result in the formation of an hydroxycarbonate apatite (HCA) to which bone can bond. Bonding occurs due to a sequence of reactions:

(a) Leaching and formation of silanols (SiOH)

(b) Loss of soluble silica and formation of silanols

(c) Polycondensation of silanols to form a hydrated silica gel

(d) Formation of an amorphous calcium phosphate layer

(e) Crystallisation of a hydroxycarbonate apatite layer

These five reaction stages do not depend on the presence of tissues. They occur in distilled water, or in any simulated body fluids. The rate of tissue bonding appears to depend on the rate of HCA formation. However, bonding to tissues requires an additional series of reactions, which are less well defined. Currently, the sequence of events that appear to be associated with the formation of a bond with tissue follows:

(a) Absorption of biological moieties in the SiO_2-HCA layer

(b) Action of macrophages

(c) Attachment of stem cells

(d) Differentiation of stem cells

(e) Generation of matrix

(f) Mineralisation of matrix

The rate of development of the interfacial bond between implant and bone can be referred to as the level of bioactivity. Hench (1988) introduced an index of bioactivity, which is given as:

$$IB = 100/t_{0.5bb}$$

where $t_{0.5bb}$ is the time for more than 50% of the implant surface to be bonded to bone.

Mitral valve replacement

Recent success with short term artificial heart valves has led to much research in the development of new long life valves. A proposed design for a porous alumina mitral valve has been suggested by Gentle (1980). Because of the bioinertness and low wear rates of alumina the material could be expected to last many years in the human body. A second advantage of the porous ceramic is that the pores will support the mechanical attachment of live cells which will protect the blood from mechanical damage by the implant. His studies in pigs have shown short term success and that anti coagulant drugs are not necessary with the implant. The reduction in drug intake greatly improve the quality of life for a recipient.

3.6 Advanced ceramic materials: Summary of possible applications

3.6.1 Introduction

Ceramics, which exhibit very specific properties and cannot be replaced by other materials, play a significant role in the performance of numerous devices and equipments. In order to tailor a device property to its function, a good knowledge of the correlation between property and microstructure is required. Because the microstructure depends on the processing route, ceramic processing methods are decisive factors in developing the ceramic market.

While current ceramic applications are well known, scientists in the field continue to develop novel processing methods that can produce components of a complex shape and high reliability with a minimum of machining and at a lower cost.

3.6.2 State of the art

The current status of research and development concerning materials and composites has recently been described. In this chapter, we focus on areas of emphasis for new materials, new applications and areas for application, and new processes.

Many new materials are worth further examination, some of which are mentioned below:

(a) Research activity in inert or active bioceramics has been significant and will have an effect in the medical and health fields, prompting new actions for the future. Abrasives for cut-off or grinding wheels and cutting tools for metalworking are important for industrial applications, and not enough research is dedicated to these materials. Therefore, hard and superhard ceramics are worth studying more extensively.

(b) Methods for studying the high-temperature behaviour of particle filters, corrosion behaviour in particular, need improvement. Decomposition of SO_x and NO_x, pollutants resulting from the combustion process, must be realized through catalytic reduction at low temperature.

(c) Ceramic membranes are very useful for liquid (water) or gas hyperfiltration at a molecular scale.

(d) Electroceramics have a wide range of applications and correspond to significant research activities in the fields of ion-conducting ceramics (such as batteries and sensors), electrical insulators (substrates and multilayer integrated circuit packages), semiconductors (sensors), and superconductors.

(e) Ferroelectric ceramics are very promising for many applications such as capacitors, sensors, piezoelectric transducers, electrooptic devices, and thermistors, giving rise to productive research activities.

(f) Ferromagnetic ceramics play a significant role in the electronics industry, and their applications are extensive, including recording, telecommunications, televisions, and electronics.

(g) Miniaturization for electroceramic components and devices must be a priority for the future.

Chemical, physical, and mechanical characterization techniques have been suitably developed for materials applications. Using these tools, we need to explore new areas of materials application, some of which are listed below:

(a) In transport, catalyst supports for purification of automobile exhaust fumes are now finalized, but particle filters for diesel engines need further investigation to take into account corrosion problems. The monolithic or fibre-composite ceramic brakes for trucks, trains, planes, and automobiles require further development in terms of lifetime, cost, and performance. The use of other components (such as valves, seals, turbocharger rotors, cam rollers, tippers, etc.) is actually limited because of cost and reliability problems. Components for gas-turbine power generators have to be improved. Major efforts are being made to develop a high-efficiency solid oxide fuel cell.

(b) For energy production, several routes using coal, gas, or nuclear fuels are being considered. The performance of coal plants are limited by the thermomechanical and chemical behaviour of refractories, and a better knowledge of this behaviour is needed. Improvement of combustion turbines requires an improvement in the behaviour of used ceramics, the use of more efficient thermal barrier coatings, and a decrease in cost. The problem of how to clean up radioactive waste from nuclear power plants must be solved. Environment factors to take into consideration include type and quantity of waste, health risks, level of pollution, and route of contamination (soil, water, air). Fibres, whiskers, chromium oxide, and small particles can be carcinogenic, and substitutes must be found.

(c) The building industry has a significant influence on people's lives and more emphasis needs to be placed on research. Materials that are important include cements, concretes, gypsum cements, roof tiles, and bricks. Traditional ceramics are used throughout industry, but their development is hindered by the lack of interest in ceramic processing technology and its role in microstructure and properties. An area of growth is ceramic heritage, an application used to preserve and restore buildings, monuments, etc., which is interesting more and more researchers.

Several processing methods are well developed, including the characterization of powders, microstructure, and porosity; others need more attention:

(a) Powder synthesis requires complementary studies, the powder behaviour being fundamental for shape-forming processes and densification. Some shape-forming processes, such as pressing, slip casting, and tape casting, are well known. Sintering has been extensively studied and there is a good understanding of the involved mechanisms for ceramics. "Modelling" density variation during pressing and shrinkage during sintering are needed in order to obtain near-net shape parts and to minimize machining, which represents a large portion of the fabrication cost.

(b) New routes for near-net shape forming of advanced ceramics and composites have been developed and hold promise for the fabrication of components with high reliability at acceptable costs.

(c) Injection moulding offers great promise for low-cost fabrication but needs more attention in the area of organic vehicle removal.

(d) Pyrolysis of inorganic polymers can be used for the production of ceramic parts, but this process results in significant shrinkage, which can lead to the introduction of flaws. Therefore, this route is promising but requires further research concerning the synthesis of inorganic polymers and the transformation of polymers to inorganic solids.

(e) Quality control methods for ceramic components are not often studied. The area of nondestructive techniques, in particular, needs more attention.

Compared to those of the U.S. and Japan, the EU's fundamental research activities have been quite effective in the areas of processing, bioceramics, ceramic matrix composites, carbon-carbon composites, cutting tools, nuclear ceramics, and electroceramics, but rather low regarding filters, building materials, refractories, fibres, gas turbines, abrasives, superhard ceramics, nanomaterials, function gradient materials, and finishing.

The EU, U.S., and Japan have different research strategies. In Japan, national projects with technological goals are defined and funded by the government. A similar approach for some programmes exists in the U.S. In contrast, European projects are rather broad with less precise goals, but there are national projects with more specific objectives.

3.6.3 Trends for the future

The socioeconomic impact of the following themes will be taken into account in establishing priority actions and basic research necessary to reach fixed objectives which are summarized in Figure 3.2 and Table 3.3.

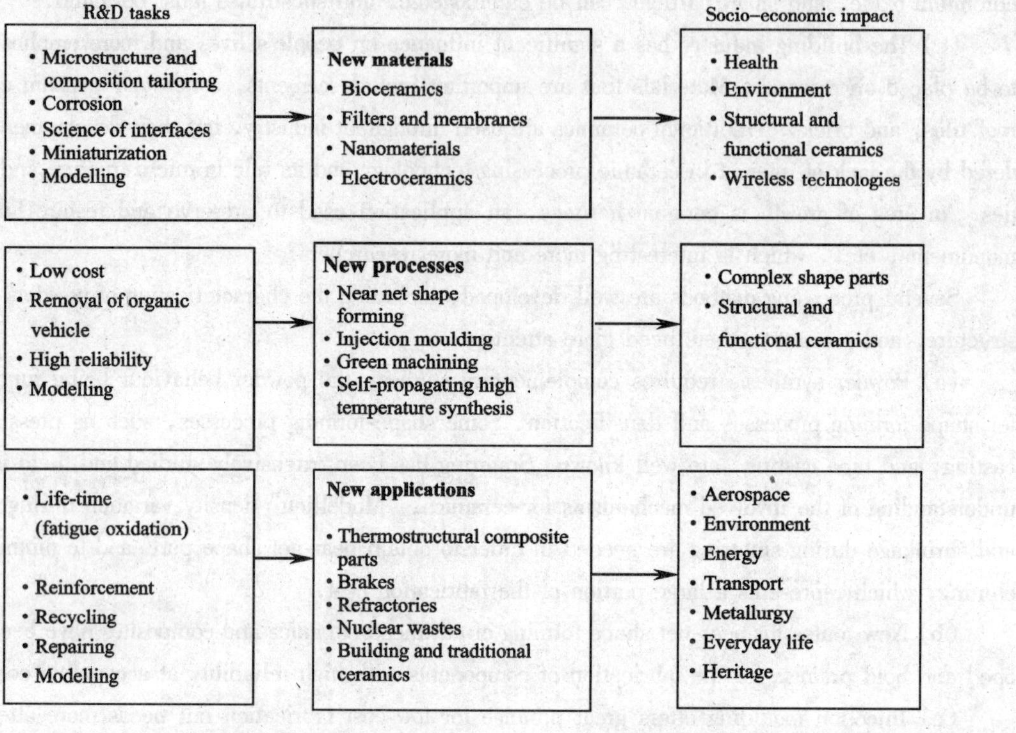

Figure 3.2 The socio-economic impact of ceramic materials

Table 3.3 Needs in basic research for ceramic materials

• Near-net shape forming	• Nanomaterials and nano-composites
• Processing of complex shapes with high reliability	• Nanostructured coatings
• Hybrid bones	• Sensors
• Catalytic reduction	• Electro-optical devices
• Membranes	• Superionic conductors
• Nuclear wastes	• Ferroelectric
• Recycling	• Miniaturization
• Antibacterial actions	• Multilayer materials
• Nanopowders	• Interfaces
• Carbon nanotubes	• Self-healing composites

New materials, devices, and applications

The following are considered priority actions:

(1) Bioceramics: Application to health

Today's orthopaedic implants have an average life expectancy of 10 to 15 years, which is generally limited by osteolysis, which is caused by the wear of polyethylene cups against metal or ceramic (i.e., alumina- or yttria-stabilized zirconia) femoral heads. However, considering our aging population, there is a growing demand for implants with a life expectancy of more than 30 years. They must include new generations of ceramic-ceramic wear couples for hips, and knee prostheses that can last where ceramics have so far failed.

Therefore, existing ceramics must be improved in terms of reliability (resistance to slow crack propagation for alumina-based materials and stability, and in particular for zirconia ceramics, which presents a problem for surface low-temperature degradation). When working on ceramic processing, scientists must strive to refine microstructure (colloidal and sol-gel) processes and increase ceramic stability and crack resistance. Ceramic processing must include the ability to make complex shapes (i.e., knee prostheses) with high reliability. We must incorporate new ceramic materials into orthopaedic design. They must address current biocompatibility problems and have properties superior to monolithic alumina and zirconia. There is a new interest in the potential for the biomedical applications of zirconia-toughened alumina. Materials scientists are already familiar with zirconia-toughened alumina, which is exceptionally hard, strong, and crack resistant. However, its superiority for joint prosthesis application must now be proved by simulator studies under physiological conditions (where there is a need for knee prostheses simulators and complex modes of fracture studies), industrial manufacturing, and in vivo preclinical evaluation.

Bioactive ceramics, mainly hydroxyapatites, are used today as coating on femoral stems and acetabular cups for noncemented hip and knee joint prostheses. However, their use must spread to other applications, such as bone substitutes. Most orthopaedic surgeons agree that synthetic bone substitutes sometimes result in clinical failure caused either by materials fracture or lack of bone integration. The challenge is therefore to tailor the microstructure and the composition (hydroxyapatite/tri-calcium phosphate composites) of bone substitutes to adapt to the biofunctionality in a given site. Materials scientists should also work with biologists and surgeons to better understand osteo-conduction mechanisms to back-up processing. The second generation of bone substitutes must be more active, incorporating medicines (e.g., antibiotics, chemotherapy, drugs) to act locally against disease. The third generation will be the hybrid bone, where the ceramic is the support to human cell culture. Basic research in bioceramics should focus on producing complex shapes with high reliability, developing nanocomposite ceramics with improved properties, and incorporating medicines into hybrid bone.

(2) **Ceramics and environment**

The reduction of pollutants, an actual necessity, requires research work in the following areas:

(a) Particle filters: The corrosion behaviour must be improved by acting on the ceramics microstructures, decreasing the glassy phase amount, or changing the type of ceramic (oxide or nonoxide ceramic). Furthermore, filter materials must show good thermomechanical behaviour.

(b) Development of a low-temperature catalytic reduction for the decomposition of SO_x and NO_x pollutants.

(c) Problems of refractories used in waste treatment plants and treatment of flying ashes.

(d) Development of new ceramic membranes for hyperfiltration and nanofiltration of liquid or gas, often very aggressive, for desalination.

(e) Nuclear waste treatment: vitrification, and cementitious materials.

(f) Ceramic recycling.

(g) Development of antibacterial tiles, new sanitary wares, or smart toilets.

Priority research in environment must be focused on catalytic reduction, filters, membranes, nuclear wastes, recycling, and antibacterial actions.

(3) **Nanomaterials and functional gradient materials**

It is expected that nanostructured ceramics play an important role in following areas:

(a) Nanocomposites are known to exhibit unusual properties. Creep resistance, and thermomechanical or wear behaviour can be improved by using nanocomposites. Nanocomposites or nanoceramics exhibit superplasticity, which can be used for shaping.

(b) Nanostructured coatings on substrates allow increased wear and corrosion resistance and durability of cutting tools. The different processes used for obtaining nanostructured materials and nanomaterial behaviour must be investigated further.

(c) Carbon nanotubes show exceptional mechanical and electronic properties, and can be very high-performance materials. This new class of materials merits greater attention.

(d) Functional gradient materials reduce thermal stresses for thermal barriers and can be used in a number of applications, including thermal insulation, thermal fatigue, oxidation and corrosion resistance, and electrical, magnetic, or optical properties. Research activity in functional gradient materials (processing and properties) must be encouraged.

(e) Superhard ceramics are interesting materials, but research in this range is insufficient.

(f) Ceramic-ceramic and ceramic-metal joining is important for the production of complex structures and requires further development.

Preferably, research in nanomaterials should focus on nanostructured coatings and carbon nanotubes.

(4) **Electroceramics**

Electroceramics are subject of broad interest to the research community. The following are some of the areas of electroceramics that merit increased attention: piezoelectric actuators; hybrid sensors; electrooptic devices; new ferrites; resistors; positive temperature coefficient materials; materials for batteries, fuel cells, and electrochemical capacitors; oxygen separation membranes; transport properties; and defect chemistry.

Research in the following areas must be developed:

(a) Nanopowders, nanoelectroceramics

(b) Decrease of the sintering temperature

(c) Multilayer structures: reduction of thickness, substitution of noble metals

(d) Decrease of the mechanical and dielectric loss at very high frequencies

(e) Development of sol-gel technology for thin layers

(f) New ceramics with improved properties

(g) New processing routes

(h) Miniaturization of components and devices for wireless technologies

In the area of electroceramics, particular attention must be devoted to sensors, actuators, electrooptic devices, superionic conductors, ferrites, nano-electroceramics, multilayer structures, and miniaturization in relation to new processing routes.

The science of interfaces is multidisciplinary and useful for a wide spectrum of applications, such as electroceramics, functional gradient materials, joining, thin films, and coatings, and therefore needs more research, particularly in the areas of miniaturization and nanomaterials.

New processes

A significant challenge in ceramics concerns the fabrication of complex-shaped components without defects (pores, inclusions, inhomogeneities, etc., which increase the risk of failure), with high reliability and low cost. Because machining is expensive, it will be important to develop and optimize near-net shape forming of ceramics, which allows minimizing of fabrication time, a primary competitive factor. Processing routes requiring more research include:

(a) Injection molding: This process needs a large amount of binders that must be removed without leaving flaws.

(b) Colloidal techniques: Direct coagulation casting, gel casting, and hydrolysis-assisted solidification, which need a high amount of additives.

(c) Solid freeform fabrication.

(d) Temperature-induced forming: Consolidation of a slurry is obtained by a temperature increase, and shaping is made by using a nonporous mold.

(e) Directed molten metal oxidation or reactive melt penetration for composites processing.

(f) Self-propagating high-temperature synthesis is an interesting, low-cost technique for making fine powders.

(g) Synthesis and processing of nanomaterials and nanocomposites are exciting areas to be developed.

The machining of sintered ceramics is time and energy consuming and can represent up to 80% of the production cost of a ceramic part. A possible route to reduce machining costs is to machine green ceramics, which can be an effective way to obtain complex ceramic parts. However, research in this field is sparse and needs to be developed. It is critical that the research community investigates the types of green properties required to withstand green machining. The role of binders must be specified, and specific mechanical tests for green compact characterization must be developed in order to find the correlation between mechanical behaviour and green machine performance.

Basic research in processing must be devoted to near-net shape forming, organic removal, SHS synthesis, and nanomaterials.

Other applications

(1) **Fibre-reinforced ceramic composites**

Ceramic matrix composites reinforced with fibres (CMCs) are devoted to high-performance thermomechanical applications without the brittle fracture behaviour typical of conventional ceramics. For the future, the main obstacle concerning these materials is still costs of both, raw materials and manufacturing. In addition, several properties (such as oxidation resistance, creep, and fatigue) have to be improved. Research must be developed in the following directions:

(a) Reinforcement: There is a need for cheaper fibres with improved creep resistance. The best SiC fibres are very expensive and exhibit extremely high creep and oxidation rates. Oxide fibres, like alumina, are limited because of very high creep rates. Carbon fibres need improved oxidation-protection solutions.

(b) Protection against oxidation: Development is needed of sophisticated multilayer matrices and multilayer surface coatings in order to promote the matrix cracks healing by formation of glass in the whole temperature range. Research in this field must be done in close relation with processing solutions. Brake materials (carbon-carbon or particular attention to the effect of the inhibitors on friction performance.

(c) Reliability of inhomogeneous, anisotropic, and porous materials: Development is needed of adapted nondestructive examination methods, specific strength and quality criteria, and new damage indicators. Despite the effective reduction in brittleness caused by fibrous reinforcement, further research is needed to improve toughness and thermal and mechanical shock resistance. This is particularly true in light of the goal of cheaper materials.

(2) **Refractories**

Refractories are largely used in industry — iron, steel, metallurgy, foundry, cement works, glass, ceramic, power plants, chemistry, etc. These materials play an important economic and strategic role, and the improvement of their performance constitutes a major challenge for the future. Traditionally, rather empirical methods were used to improve refractories' behaviour, but it is now very desirable to forecast and model microstructures and properties of refractories.

We must encourage the study of the microstructure-property relationship, modelling of the thermomechanical response, and corrosion behaviour and apply our results to real components.

(3) **Building materials**

Because the use of cement requires water consumption, the availability of water is critical. It is therefore necessary to improve the workability of concrete in order to reduce water consumption. Macrodefect-free cements with organic additives have more strength and lower porosity than conventional concrete. Cement strength can be increased by fibre reinforcement. Recycling fly ashes and other light aggregates in concrete must be encouraged. The ability to use cement for repair or reconstruction now needs to be taken into account. The temperature behaviour of cement and concrete also needs to be studied and improved for public safety in tunnels, buildings, and other public structures.

Research on cement, concrete, and plaster must receive increased attention. Specifically, the following areas need basic research: self-healing composites, fibres, interfaces corrosion, recycling, and new concretes.

In conclusion, new developments in ceramics require both basic and applied research in materials and properties, processing and microstructure, and design and modelling of materials and devices. The trends for the future are summarized in Table 3.3.

3.7 Advanced ceramic materials: Basic research viewpoint

3.7.1 Introduction

Ceramics are a class of materials broadly defined as "inorganic, nonmetallic solids". They have the largest range of functions of all known materials. Despite the already existing variety of compounds, the number of processing techniques, and the known diversity of properties and applications of the materials, all advanced countries share the need for basic research in several areas: finding new compounds with improved specific properties, increasing knowledge of fabrication processes for economical and ecological ceramic parts production, miniaturization and integration of ceramics with similar or dissimilar materials, and building a better understanding of materials behaviour through computational modelling.

3.7.2 State of the art

The last decades have seen the development of the enormous potential of functional ceramics based on unique dielectric, ferroelectric, piezoelectric, pyroelectric, ferromagnetic, magnetoresistive, ionical, electronical, superconducting, electrooptical, and gas-sensing properties. Such properties now constitute the basis of a broad field of applications (Figure 3.3). Scientific advances concerning many ceramic materials have enabled technological breakthroughs of truly global proportions.

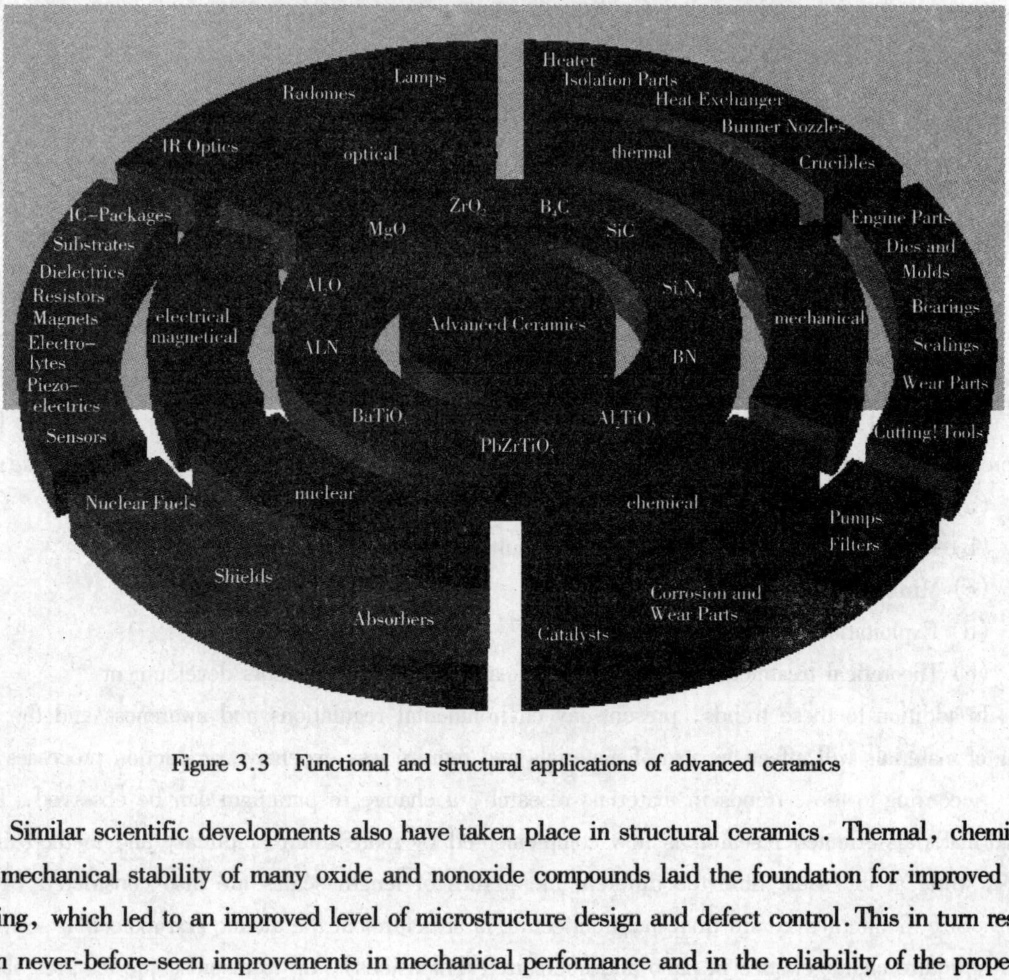

Figure 3.3 Functional and structural applications of advanced ceramics

Similar scientific developments also have taken place in structural ceramics. Thermal, chemical, and mechanical stability of many oxide and nonoxide compounds laid the foundation for improved processing, which led to an improved level of microstructure design and defect control. This in turn resulted in never-before-seen improvements in mechanical performance and in the reliability of the properties of components and devices.

In addition, superior combinations of thermal, insulating, and mechanical properties have become the basis of huge applications in the packaging of microelectronics and power semiconductors. Therefore, ceramic materials have now become the cornerstone of such advanced technologies as energy transformation, storage and supply, information technology, transportation systems, medical technology, and manufacturing technology (Figure 3.4). Since technical and economical progress in such areas is in line with the mastering of materials, empirical fabrication procedures are progressively being replaced by knowledge-based production and engineering. Therefore, today's advanced materials synthesis and processing, and the economical manufacturing of components and devices are being driven more and more by an increased understanding of the underlying fundamental scientific phenomena.

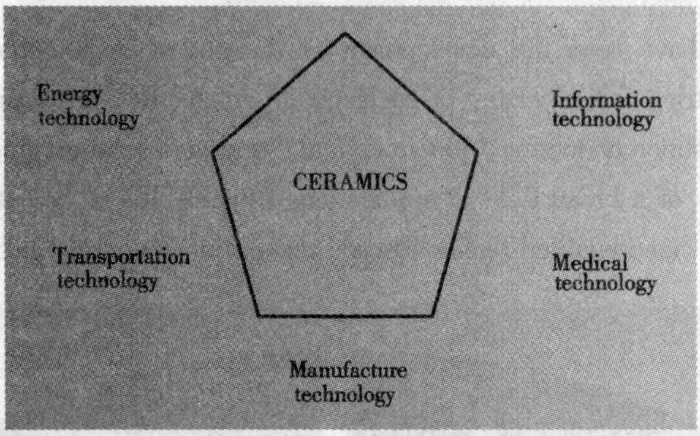

Figure 3.4 Technology areas which benefit from advanced ceramics

3.7.3 Trends in technology

Since production and engineering of materials and components are increasingly knowledge-based, the technology itself reveals gaps in the basic understanding of materials. Such gaps limit the degree to which materials technology can be competitive with other technologies. Therefore, the continued pursuit of additional knowledge will always be necessary. Some technological trends are important to follow as present and future directions of basic and applied research. The most significant of these are:

(a) Increasing materials complexity because of increased functionality

(b) Integration of different materials into multifunctional components

(c) Miniaturization of devices

(d) Exploitation of nanosized effects

(e) Theoretical treatment and modelling of materials and components development

In addition to these trends, present-day environmental regulations and awareness and the recycling of materials will affect the use of materials and require less expensive production processes.

According to these trends in materials research, a change in paradigm can be observed. Traditional materials-oriented research is now complemented by more interdisciplinary and method-linking work, while at the same time the different hierarchies of length scales are also considered (Figure 3.5). Size hierarchies extend from the architecture of structures at the atomic and molecular scale (via microstructure characteristics in nano-and microsized dimensions), up to macroscopic features of components and devices. One fundamental area of research to pursue is the development of a unified description of deformation and fracture in order to design microstructures for improved mechanical reliability of ceramic parts. This would decrease the material's susceptibility to environmental stresses and facilitate integration of ceramics into whole engineered systems. In addition, as these tasks become more and more interdisciplinary, advances in ceramic materials will benefit increasingly from closer cooperation among materials scientists, physicists, chemists, engineers, and, especially, biologists.

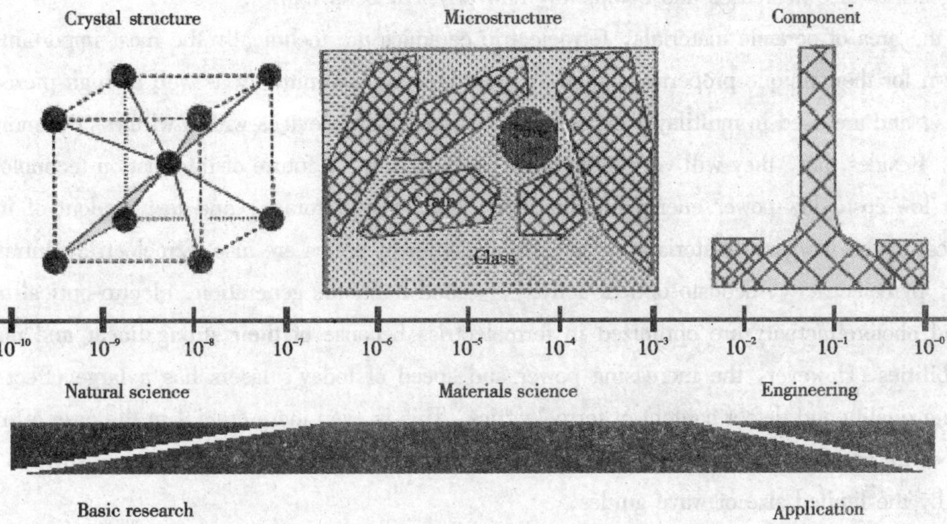

Figure 3.5 Hierarchies of length scale and multidisciplinary character of materials

3.7.4 Needs for future basic research

Following technological trends, the needs for future basic research in the field of ceramics can be divided into four major areas: ① materials and materials properties research in order to widen the area's scope and match its needs for future applications, ② research to increase the knowledge of economical and ecological production processes for materials, components, and devices, ③ miniaturization and integration, and ④ modelling and numerical simulation, which would complement or even act as a substitute for present areas of experimental work, thus not only directing research to defined questions, but also reducing practical work and time periods typically combined with product development. The identified four areas of future basic research will be discussed in the following.

Materials and their properties

Ceramics now cover an extensive area of functionalities. Hardly any other class of materials offers such a variety of properties useful for applications. Promising technical developments emerge from the discovery of new compounds, and new materials with specific functions have recently appeared on the horizon. These materials include high temperature superconductors, ultra high-strain single crystal piezoelectrics, high-frequency dielectrics, phosphors, intercalation compounds, ductile layer-structured ceramics, and thermoelectrics. Detailed basic research in solid state chemistry and solid state physics (i.e., activities focused on the elaboration of new compounds and the understanding of their basic properties) will continue to widen the scope of materials, resulting in the optimization of existing materials' functionalities and the likely enhancement of ceramics with new physical behaviours. Since both result in complex materials, there will be a continuous need for the comprehensive evaluation of the underlying materials (i.e., structure analysis and phase studies of multicomponent systems). Therefore, extended investigations of physical, chemical, and mechanical properties will be necessary in order to evaluate the potential of the materials for practical applications. Of special interest will be the investigation of nanosized effects onto certain properties (i.e., continuous or discontinuous property changes with size decreases and completely new physical behaviours).

In the area of ceramic materials, ferroelectric ceramics are technically the most important. They are known for their unique properties, such as high dielectric permittivity as well as high piezoelectric constants, and are used in multilayer capacitors or as microwave devices within wireless communication systems. Besides that, they will very likely play a key role in the future of information technology as a basis for low-cost, low-power energy consumption, highdensity storage, and fast readout of information. Most of the inorganic materials for linear and nonlinear optics are also ferroelectric. Infrared absorption, pyroelectricity, acousto-optical activity, second-harmonic generation, electro-optical modulation, and photorefractivity are optimized in ferroelectrics because of their strong linear and nonlinear polarizabilities. However, the increasing power and speed of today's lasers has a large effect on the crystalline quality and defect content of ferroelectrics. This is even more crucial in the case of integrated devices for which optical density, electric fields, thermal gradients, and mechanical stresses are increased by the limited size of wave guides.

Presently, piezoelectric ceramics certainly show the broadest range of individual applications, including sonars, ultrasonic cleaners, buzzers, accelerometers, hydrophones, surface acoustic wave filters, delay lines, piezotransformers, ultra-precision positioners, and many other sensor and actuator functions requiring a wide field of interdisciplinary research for further evaluation.

Magnetoresistive materials, which change their electrical resistivity upon application of magnetic fields, have also received attention for magnetic reading heads. Ceramics, such as doped rare-earth manganites (in which so-called phenomena of colossal magnetoresistance were observed) exhibit extremely high values for this effect, even superior to that of giant magnetoresistant metals presently in use.

Ion conduction also provides great technological benefits. It is a basis for energy and information technology systems such as batteries, fuel cells, chemical sensors, and chemical filters providing efficient and cleaner energy transformation, chemical control, and environmental protection. Apart from the functional aspect, the application of these materials requires the simultaneous fulfilment of many properties. These include thermal stability, chemical stability, mechanical stability, and electrical stability. Examples in which fundamentals led to new materials and devices are stable proton conductors, interfacially controlled ionic conductors, and chemical sensors for acid/base active gases.

In the field of high-temperature superconductivity, important progress has been achieved over the last few years regarding both energy applications as well as electronics. Large-sized demonstration prototypes of superconducting cables, transformers, and fault-current limiters have been fabricated and successfully tested. Some products, such as current leads for low-temperature systems and filters for mobile telephone base stations, have already been introduced into the market in a competitive way.

Characterized by a steady increase in energy density imposed by phenomenal requirements in the thermal management of last-generation integrated semiconductor devices, highly thermal conductive materials represent a more important class of functional ceramic materials.

There is much room for development in basic research, not only in functional materials, but also in structural materials. Superhard materials for cutting tools, abrasives, and tribological applications, as well as thermal barrier coatings to improve gas-turbine efficiency, and fibre-reinforced ceramics for failure-tolerant high-temperature components, are just a few examples of great technological importance.

A unique set of properties is exhibited by a newly identified family of ternary-layered hexagonal carbides and nitrides (e.g., Ti_3SiC_2). These ceramic compounds are malleable at room temperature and deform plastically at elevated temperatures. They warrant further research.

Other important classes of ceramics to be studied both for functional as well as structural applications are negative and low-thermal expansion materials such as beta eucryptite, sodium zirconium phosphates, vanadates, and tungstates. There is a widespread need for these materials in applications that require either dimensional stability or thermal-shock resistance as well as in electronic devices, mechanical parts, catalyst supports for automotive applications, high-performance optical mirror substrates, ceramic hosts for nuclear waste immobilization, port liners for diesel engines, and coating materials for C/C composites.

Processing and microstructural design

Ceramics are synthesized into glasses, polycrystals, and single crystals, and many forms dictated by their use, including fine powders, fibres, thin films, thick films, coatings, monoliths, and composites. Polycrystalline components are conventionally produced by powder synthesis and forming processes followed by sintering at high temperature. The performances of ceramic materials are determined not only by the structure and composition, but also by defects (such as pores), second phases (which can be deliberately added to facilitate processing), and interfaces. Thus, one of the primary limitations of current ceramic processing technologies is that they are an art as well a science. This situation is changing progressively. For instance, in order to meet the requirements of miniaturization and integration, these techniques have to be supplemented by deposition techniques, such as physical vapour deposition (PVD), pulsed-laser deposition (PLD), chemical solution deposition (CSD), and chemical vapour deposition (CVD). During these processes, the materials are synthesized on a microscopic scale without powder processing as an intermediate step, usually at temperatures much below the typical sintering temperature of bulk materials. Not only can ceramic materials of small dimensions (e.g., the size of semiconductor chips) then be deposited, thus responding to the integration needs of different materials, but their ceramic compositions and structures can also be designed much more flexibly. The understanding and mastering of phenomena at the microscopic and, more importantly, at the nanoscopic scales clearly need further basic research. A specific example can be found in thin films of ferroelectrics, where dielectric properties are dominated by interfacial effects rather than by the bulk capacitance.

A great challenge is the architecture of ceramic materials in atomic dimensions, not only in thin films but also in thick films and bulk materials. The ceramic materials will open up novel classes of materials with properties still unknown with respect to conventionally processed materials. One possibility for research is the thermally induced transformation of preceramic compounds by solid state thermolysis (SST). By tailoring the composition and molecular structure of precursors by means of advanced chemical syntheses, and by controlling thermolysis, the composition, structure, and microstructure of ceramic materials can be designed. The characteristics of precursor thermolysis, and the relative ease with which various geometries can be processed at the preceramic stage (such geometries include fibres, thin films and coatings, infiltrates, bulk materials, etc.), using standard polymer processing techniques, makes them highly applicable for the production of fibre-reinforced composites, oxidation-resistant coatings, wear-resistant materials, and many others.

Another growing field of research with an attractive technical potential is the production of materials inspired by biomineralization. Oxidic deposits from aqueous solutions on organically modified surfaces, for example, provide a synthesis route for ceramics at ambient conditions. Such template-induced and self-assembled materials using biomimetic techniques provide "soft", low-cost routes, offering real opportunities to get new metastable materials that possess original microstructures as layered polymer-ceramic composites and cannot be obtained by conventional high-temperature routes. Therefore, the investigation of interactions between organic and inorganic phases, as well as nucleation and growth phenomena, is a promising area of basic research.

Nevertheless, conventional elaboration processes will continue to offer breakthroughs. A deeper understanding of the surface chemistry of powders will lead to a better control of the interaction between particles during colloidal powder processing, thus providing a means for controlling powder consolidation and defect-free microstructure. New methods for net-shape forming based on the control of interparticle forces in dense suspensions of powders have been proposed recently. There is no doubt that possibilities offered by surface chemistry are far from being fully exploited, especially in the case of nonoxide ceramics. Among the challenges for future years is the development of processing paradigms for the production of finer powders, which could lead to new applications resulting from different green-state architectures.

Direct and indirect cost savings by process-cost reductions and product improvements, respectively, are necessary in the field of advanced properties of ceramics. Cost savings are typically achieved by a better understanding of the fundamentals of the underlying processes. Savings could also be obtained by new types of low-cost processes, component design changes or simplifications by using proper joining techniques, or product-development time reductions (e.g., by rapid prototyping). In essence, processing science remains a net priority, but in the case of ceramics, it now extends far beyond the simple processing of powders.

Miniaturization and integration

One of the great challenges and opportunities in advanced processing lies in the miniaturization and integration of different classes of materials (e.g., ceramics, metallic alloys, polymers). Miniaturization requires a search for new processing techniques adequate for the manufacturing of miniaturized components and devices, but also a proper means to control microstructure at the mesoscopic level. A new field of research emerges as many characteristics reveal side effects at the mesoscopic dimension. For instance, the properties of thin films and nanostructured materials often deviate from those of corresponding microstructured bulk materials.

Due to miniaturization, nanosized effects and nanotechnologies will thus gain importance. The miniaturized devices may include ultrathin films, multilayers, interface controlled materials, nanocomposites, metastable systems with high information content on the nanoscale, and lateral nanostructuring. Therefore, miniaturization provides an innovative means for system development.

Furthermore, present challenges to technological development often lie in engineering the miniaturized devices. Integration of different ceramic materials with metals or future polymers is one way to miniaturize systems and to produce one "function" for both functional and structural applications. Integration will thus play an increasing role in many technologies. In the case of microelectronics, for example, driving forces are the integration of various functions into conventional semiconductor chips and the evolution of multifunctional components and systems. The area of integration would benefit from a clear understanding of effects that occur during the co-production of different materials in a single device, and also of the role played by interfacial chemistry in adhesion and fracture. Concerning microelectromechanical systems (MEMS) all scientific background of mechanical behaviour must be revisited for two reasons: ① classical fracture mechanics may not be appropriate at the microscale when the size of the component approaches grain size; and ② the validation of the design will require new testing methodologies for strength, toughness, fatigue resistance, etc.

Modelling and numerical simulation

Modelling and simulating have been used for a long time in the development and improvement of materials. On one hand, atomic dimensions have been successfully described by, for example, first principle and *ab initio* quantum concepts. Fundamental theories exist for many physical phenomena (e.g. elasticity, plasticity, fracture, thermodynamics, electricity, magnetism), all of which improved enormously the understanding of mate rials. On the other hand, there have been developed a variety of chemical continuum concepts and macroscopic finite element methods to simulate the behaviour of materials and compounds. Both areas have created much progress in materials science and technology. However, in order to achieve a real breakthrough in designing new materials and tailoring their properties, there is a need to bridge both areas. The most desirable breakthrough would be to outline the structural and functional performance of a material directly from its elemental composition, structure, and morphology (Figure 3.6). Modelling concepts, which cover a complete sequence from the state of matter (via material synthesis and processing) to a component's properties and performance, would be the most desirable in materials science and engineering.

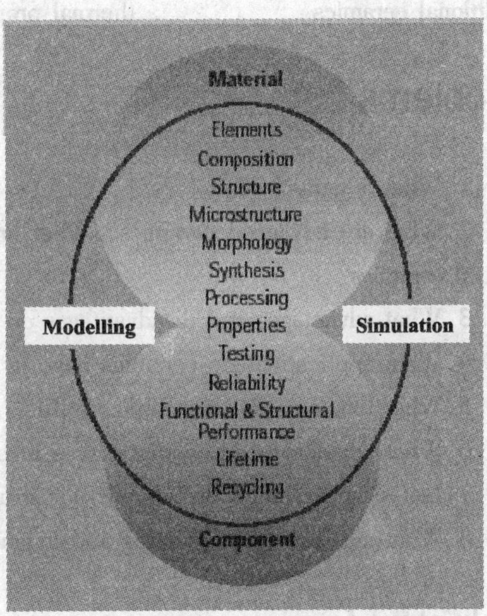

Figure 3.6 Modelling and simulation in material science and engineering

Keywords

advanced ceramics
artificial joints
alumina ceramics
borides
ball bearings
bioceramics
bioinert ceramics
bioactive ceramics
biocompatibility
bioglass
ceramics
corrosion-resistance

carbides	ceramic matrix composite	cutting tool
electrical property	engine components	electronic ceramics
functional ceramics	filters	fuel cells
glass ceramics	grinding tool	inorganic
implants	integration	laser crystals
magnetic property	mechanical property	magnetic ceramics
microstructural design	miniaturization	modelling
nonmetallic	nitrides	nanotechnology
numerical simulation	optical property	oxidation-resistance
oxides	structural ceramics	silicates
superconductor	sensor	solid-state batteries
traditional ceramics	thermal property	zirconium-dioxide ceramics

Problems

1. What is ceramics?

2. What are advanced ceramics? What are the differences between traditional ceramics and advanced ceramics?

3. What advantages and disadvantages do ceramics have over other materials?

4. What applications do ceramics have in electronic fields?

5. What functions do bioceramics fulfil?

6. What advantages do ceramics have in biomaterial application?

7. List some examples of ceramics for structural application.

8. What are main strengthening and toughening mechanisms in structural ceramic material?

Selected from

[1] The American Ceramic Society: About Ceramics. http://www.acers.org/acers/about ceramics.asp

[2] http://secure.britannica.com/ebc/article-9117301. Advanced ceramics

[3] G. Geiger. Introduction to ceramics. http://www.newi.ac.uk/buckleyc/ceramics.htm

[4] http://www.foresight.gov.uk/Previous _ Rounds/Foresight _ 1999 _ 2002/Materials

[5] D. R. Clarke, Sheik A. Akbar, I-Wei Chen, et al. Fundamental Research Needs in Ceramics. http://www-unix.ecs.umass.edu./~jakus/nsf./nsf.ceramics.report.web.pdf

[6] http://www.npl.co.uk/materials/. What are electroceramics.

[7] A. Berndt. Introduction to Bioceramics. http://www.mmat.ubc.ca/courses/mmat382.

[8] G. Fantozzi, D. Rouby, J. Chevalier and P. Reynaud. European White Book on Fundamental Research in Materials Science. 2000.

[9] F. Aldinger, J.F. Baumard. European White Book on Fundamental Research in Materials Science. 2000.

4 Polymer Materials

4.1 Introduction

4.1.1 Formation

Not all the substances can be polymerized. It is necessary that a chemical substance can act as monomer only if it has at least bifunctionality. The bifunctionality may result due to the presence of at least a double bond (like in olefins) or two reactive functional groups (two – OH groups in ethylene glycol) as follows.

$$H_2C{=}CH_2 \longrightarrow \text{+}CH_2{-}CH_2\text{+}_n$$
ethylene polyethylene

$$HO{-}CH_2{-}CH_2{-}OH \longrightarrow \text{+}CH_2{-}CH_2{-}O\text{+}_n$$
ethylene glycol poly(ethylene glycol)

Consider the hydrocarbon ethylene ($CH_2{=}CH_2$), which is a gas at the ambient temperature and pressure. If the ethylene gas is subjected catalytically to appropriate conditions of temperature and pressure, it will transform into polyethylene (PE), which is a solid polymeric material. This process begins when an active monomer is formed by the reaction between an initiator (a radical, R·) or catalyst species and the ethylene monomer unit, as follows:

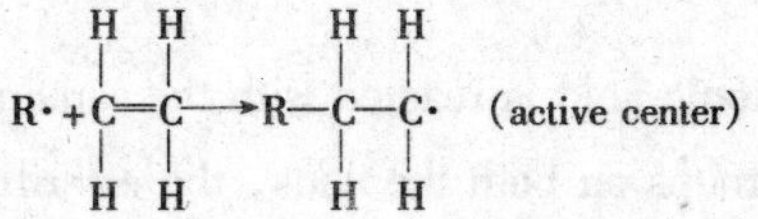

The polymer chain then forms by the sequential addition of the ethylene monomer units to this active center. The active site, or unpaired electron (denoted by ·), is transferred to each successive end monomer as it is linked to the chain. The final result, after the addition of many ethylene monomer units, is the polyethylene molecule, a portion of which is shown in Figure 4.1(a). This representation is not strictly correct in that the angle between the singly bonded carbon atoms is not 180° as shown, but rather close to 109°. A more accurate three-dimensional model is one in which the carbon atoms form a zigzag pattern (Figure 4.1b), the C — C bond length being 0.154 nm. In this discussion, depiction of polymer molecules is frequently simplified using the linear chain model.

Now we consider a reaction between two compounds each containing two reactive functional groups such as ethylene glycol with two hydroxyl groups and malonic acid with two carboxyl groups.

In the first step, the reaction between the two compounds gives rise to a monoester. Then, the

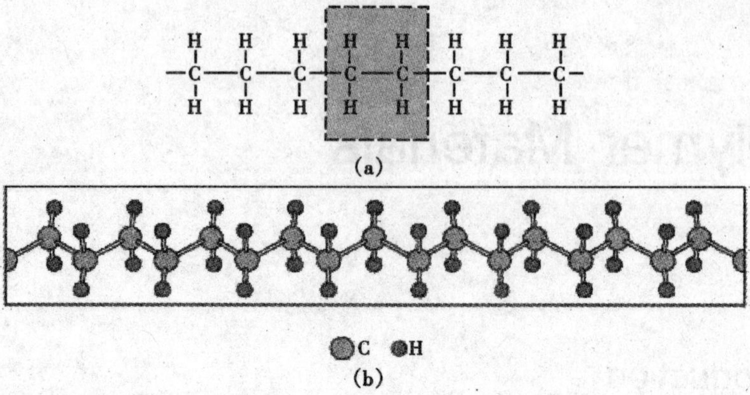

Figure 4.1 For polyethylene, (a) a schematic representation of monomer and chain structures, and (b) a perspective of the molecule, indicating the zigzag backbone structure

monoester can combine with either the glycol or the malonic acid at the corresponding functional group end to facilitate further growth of the chain by forming a diester.

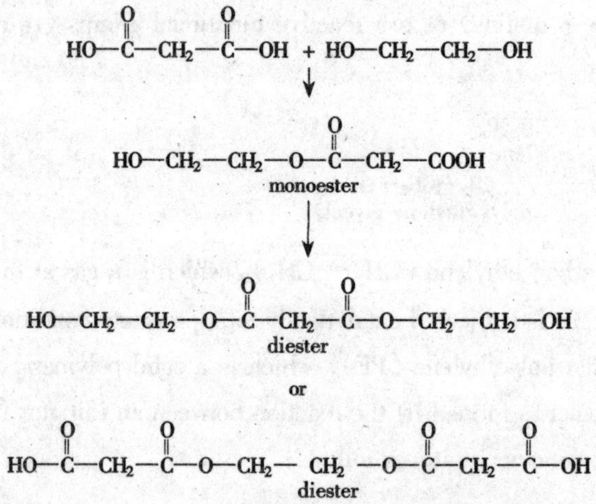

Thus if n moles of a bifunctional species such as malonic acid is reacted with the same number of moles of ethylene glycol which has two reactive hydroxyl groups on both the ends, the esterification reaction occurs stepwise forming mono, di-, tri-, tetra-, multi- and polyesters as following.

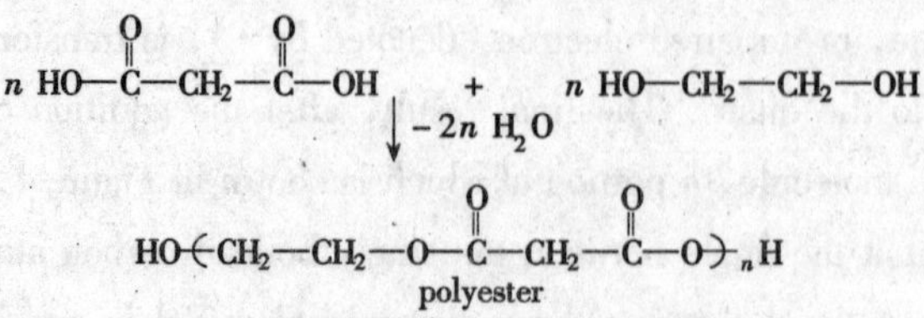

It can be inferred that the condensation reaction of each of the monomer to another continues till the concentration of one of the species is depleted. Then one can have high molecular weight polymers formed from these bifunctional monomers. Thus, the reaction between two bifunctional molecules leads to a linear high polymer. Further, if we add some trifuctional group, it may form either branched polymeric chain or a three-dimensional network structure.

4.1.2 Classification

There are several ways to classify polymers, the classification is based on several considerations, the source of polymers i.e. natural or synthetic, the type of polymerization process used in the synthesis, nature and type of chain and solid state behavior of polymer chains etc. The different classification schemes can be outlined as shown in Table 4.1.

Table 4.1 Polymer Classification

Methods	Polymer types
Sources	Natural and synthetic polymers
Type of polymerization	Addition and condensation polymers
Structure and composition	Homopolymers and copolymers
	Homochain and heterochain polymers
	Organic, elementoorganic and inorganic polymers
	Crystalline and amorphous polymers
	Linear, branched and crosslinked polymers
Applications	Thermoplastics and thermosets
	Fiber, plastics and elastomers
	General polymers, functional polymers and biomedical polymers

Polymers can be classified as natural or synthetic. The common natural polymers include polysaccharides (starch, cellulose, gums etc.), proteins (gelatin, albumin, enzymes, insulin), polyisoprenes (natural rubber, gutta percha) and nucleic acids (RNA and DNA). Natural polymers are sometimes also called "biopolymers" or "biological macromolecules".

According to the structure of their main chain, all polymers can be broadly classified as homochain and heterochain polymers. In case of homochain polymers the main chain is made up of linkages between the atoms of same element e.g. carbon in organic polymers. In heterochain polymers, the main chain consists of linkages between different atoms e.g. between carbon and oxygen in polyester, polyethers, carbon, nitrogen and oxygen in polyurethanes etc.

According to the response of a polymer to mechanical forces at elevated temperatures, thermoplasts (or thermoplastic polymers) and thermosets (or thermosetting polymers) are the two subdivisions. Thermoplasts soften when heated (and eventually liquefy) and harden when cooled. And, these processes are totally reversible and may be repeated. On a molecular level, as the temperature is raised, secondary bonding forces are diminished (by increased molecular motion) so that the relative movement of adjacent chains is facilitated when a stress is applied. Irreversible degradation results when the temperature of a molten thermoplastic polymer is raised to the point at which molecular vibrations become violent enough to break the primary covalent bonds. In addition, thermoplasts are relatively soft. Linear polymers and those having some branched structures with flexible chains are thermoplastic. These materials are normally fabricated by the simultaneous application of heat and pressure. Thermosetting polymers become permanently hard when heat is applied and do not soften upon subsequent heating. During the initial heat treatment, covalent crosslinks are formed between adjacent molecular chains; these bonds anchor the chains together to resist the vibrational and rotational chain motions at high temperatures. Crosslinking is usually extensive, in that 10 to 50% of the monomeric units are crosslinked. Only heating to excessive temperatures will cause severance of these crosslink bonds and polymer degradation. Thermoset polymers are generally harder and stronger than thermoplastics, and have better dimensional stability. Most of the crosslinked and network polymers, which include vulcanized rubbers, epoxies, and phenolic and some polyester resins, are thermosetting.

Polymers can also be classified further as fibers, plastics, resins and rubbers based on the nature and extent of secondary valence forces and mobility among the constitutional repeat units. Important examples belonging to each class are shown in Table 4.2. Plastics, fibers and elastomers possess different physical properties. While fibers are crystalline materials, elastomers are amorphous.

Table 4.2 Some important polymeric materials useful as plastics, fibers, rubbers and resins

Form	Type	Polymeric materials
Fibers	Natural	Cotton, wool, silk and asbestos
	Semi-synthetic	Viscose rayon, acetate rayon and cuprammonium rayon
	Synthetic	Polyamides, polyesters, acrylics, polyurethanes, polypropylene, poly(vinyl chloride), poly(vinyl alcohol)
Plastics	Cellulosic plastics	Celluloid, cellophane
	Vinyl plastics	Polyethylene, polypropylene, polytetrafluoroethylene (Teflon), polystyrene, poly(vinyl chloride), poly(vinyl acetate)
	Polyacryloids	Poly(acrylic acid), poly(methyl methacrylate)
Rubbers	Natural	Polyisoprene
	Synthetic	Polybutadiene, polychloroprene, butadiene copolymers (e.g. NBR, SBR), butyl rubber, polyurethanes, polysulfides, polysiloxanes (silicone rubber)
Resins	Water soluble	Modified starches, cellulose derivatives, poly(vinyl alcohol), poly(vinyl pyrrolidone), poly(acrylic acid), polyacrylamide
	Water insoluble	Phenol plasts and amino plasts, polyurethane resins, silicone resins, epoxy resins, unsaturated polyesters

4.1.3 Nomenclature

Polymers are prepared by addition polymerization of an unsaturated monomer or by step polymerization of monomer (or monomer pair) containing at least two functional groups. The nomenclature of addition polymers is relatively simple. Condensation polymers, for example, terylene and nylons are named according to the repeat unit.

Addition polymers are usually named by prefixing "poly" before the monomer from which the polymer is derived. For example poly(vinyl chloride) and polyethylene are synthesized from vinyl chloride and ethylene monomers, respectively. A wide diversity exists in naming polymers viz. based on source, structure, trade names, and even IUPAC names. The nomenclature of a few polymers are illustrated in Table 4.3. However, no single nomenclature is universally adopted. IUPAC names of some other common polymers are polystyrene (poly(1-phenylethene)), polyphenylene oxide (poly(oxy-1,4-phenylene)) and polyvinyl chloride (poly(1-chloroethene)).

Copolymers are named by inserting *-co-* (or *-a-*, *-r-*, *-b-*, or *-g-* if the position of the two monomers is known). For example, poly(styrene-*co*-butadiene) is a copolymer of styrene and butadiene and poly(styrene-*b*-butadiene) is a diblock copolymer of styrene and butadiene. It is also a common practice to designate a diblock copolymer as styrene-butadiene copolymer and a triblock copolymer as styrene-butadiene-styrene block copolymer. An AB block copolymer is also designatedas poly(A-*b*-B) and an ABA copolymer as poly(A-*b*-B-*b*-A). The block sequence arrangement of a dibiock copolymer is also represented by a long dash such as A-*block*-B and the corresponding copolymer as polyA-*block*-polyB.

Table 4.3 Different nomenclature schemes for polymers

Sources	Repeat unit	Name based on chemical structure	IUPAC name	Trade name
$CH_2=CH_2$	$—CH_2—CH_2—$	Polyethylene	Polymethylene	Polyox
Ethylene oxide or ethylene glycol	$—CH_2—CH_2—O—$	Poly(ethylene oxide) or poly(ethylene glycol)	Polyoxyethylene	Carbowax
–	$—CH_2—CH(OH)—$	Poly(vinyl alcohol)	–	Vinylon
Caprolactam or ε-aminocaproic acid	$—NH—C(=O)—(CH_2)_5$	Poly(ε-caprolactam)	Poly[imino(1-oxy-hexamethylene)]	Nylon 6
Ethylene glycol and terephthalic acid	$—O—CH_2—CH_2—O—C(=O)—C_6H_4—C(=O)—$	Poly(ethylene terephthalate)	Poly(oxyethylene terephthaloyl)	Mylar (film) Terylene or Dacron (Fiber)

4.1.4 History

Polymers existed in nature ever since the life originated on earth as the prime biomolecules viz. proteins, nucleic acids and polysaccharides, which are all naturally occurring polymers (biopolymers or biological macromolecules). Staudinger's concept of macromolecules (severely criticized initially) got support from the research works of Hermann Mark and Wallace Carothers. The development of molecular weight determination methods proved in no uncertain terms that molecules with longer chains and bigger size do exist.

Natural rubber (which is a polymer of isoprene) was known to man for more than 500 years. Christopher Columbus on his voyage to explore the world, found Brazilians playing with a ball-like material made from the sap (or latex) of the rubber trees. However, first successful application of natural rubber was possible only after 1840 when Charles Goodyear crosslinked natural rubber with elemental sulfur (the process is known as vulcanization). Unvulcanized rubber was a gummy material which creeped of its own weight. Different amounts of sulfur added to natural rubber made it more and more harder, tough and elastic material. Baelteland made the first synthetic polymer by the condensation of phenol and formaldehyde (now known as PF resin or Bakelite). The accidental discovery of cellulose nitrate by Schonbein and its plasticization with camphor to produce celluloid by Hyatt was done much before the concept of macromolecules was put forward by Staudinger in 1920. Some important dates in the polymer history are outlined in Table 4.4.

In the decade of 1920s several new polymers like plasticized PVC, urea-formaldehyde (UF) resins, thiokol rubber and alkyed resins were made. Polyethylene, polystyrene, polymethyl methacrylate, neoprene rubber, nylon 66, melamine resin and polyesters were soon commercialized. In 1940, Plunkett at DuPont company in USA accidentally discovered polytetrafluoroethylene (PTFE, teflon) which is most inert polymer commonly known for its use as a coating material for kitchen ware particularly in non-sticking frying pans. However, a very important development in the history of synthetic polymers is the Zieder-Natta polymerization in 1950s. Karl Ziegler was able to produce linear polyethylene (now called HDPE) under ordinary conditions of pressure and temperature. The demands of modem day life for high performing and speciality polymers has thrown a challenge to polymer chemists, physicists and technologists to come up with new polymers and also do modifications on existing polymers.

Table 4.4 Important events in the polymer history

Period	Event
1840 — 1920	Vulcanisation of natural rubber by Charles Goodyear, Schonbein made cellulose nitrate, Menard and Hyatt synthesised collodion and celluloid respectively. Chardonnet prepared regenerated cellulose fibers, Graham classified substances as crystalloids and colloids, oligomers of ethylene glycol and amino acids were made, Baekeland prepared the first synthetic polymer bakelite.
1920	Staudinger gave the concept of macromolecules, alkyd resins, plasticised PVA, UF and thiokol resins were synthesized.
1930	Synthesis and commercialisation of melamina resins, polymethyl methacrylate (Perspex, ICI; lucite, DuPont), neoprene, polystyrene, polyethylene and nylon 66 (by Carothers at DuPont).
1940	PET and unsaturated polyesters (Terylene, ICI; dacron, DuPont) and teflon (by Plunkett at DuPont) were patented.
1950	Synthesis and development of ABS, spandex (polyurethane elastomers), HDPE (by Ziegler), stereoregular PP (by Natta) and polycarbonates were reported.
1960	Ionomeric thermoplastic elastomers, aramides (Kevlar), flame retardant polymers and carbon fibers (cyclised, dehydrogenated PAN) were developed.
1970	Thermotropic liquid crystalline polymers were invented.
1980	Polyetherether ketone (PEEK) was made.
1990	Development of polymer composites, high performance polymers, design and synthesis of biodegradable polymeric vehicles for sustained release of pharmaceuticals and fertiliser formulations, design of hydro- and micro- gels based on water soluble polymers, development of conducting polymers, polymer solid electrolyte, electro-luminescent polymers and synthesis of multifunctional star and dendrimers.

In earlier days, very drastic conditions were needed for the polymerization of ethylene, which yielded branched polymer (LDPE). Both HDPE and LDPE are now important polymers with varied applications. Using Ziegler's co-ordination catalysts, Italian scientist Gildio Natta polymerized propylene with ordered arrangement of pendant methyl groups. This gave highly crystalline polypropylene, again a polymer with a great demand in today's market. Since 1980s, so many monomers were polymerized. Copolymerization using two or more monomers was achieved. The discovery of thermoplastic elastomers, thermally stable polymers, conducting polymers and liquid crystalline polymers can be considered as important landmarks in the polymer history. Polymer chemistry has shown the fastest development in past half century and has emerged as a full-fledged science having its roots in physics, biology, engineering and technology. Detailed mathematical treatments have been provided to explain the mechanisms/kinetics of polymerization and for the solid and solution properties of polymers.

4.2 Natural polymers

Living things exist and reproduce because they contain macromolecules (polymeric materials). Throughout the subsequent evolution of living organisms, polymers have been used as protective surfaces, membranes, energy storage systems, skeletal system, and pathways for electrical conduction and for countless of other purposes. These biological/natural polymers are widely known as polysaccharides, proteins, nucleic acids and polyisoprenes etc.

4.2.1 Polysaccharides

They have macromolecular structure built up of many monosaccharide units joined together with an elimination of a water molecule. Hence, polysaccharides can be broken down on hydrolysis to form monosaccharides. Polysaccharides are of almost universal occurrence in living organisms where they perform a variety of functions. In plant cells, they serve as excellent materials for construction work e.g. cell wall formation. They also serve as food materials for the nutrition of cell. Cellulose and starch are amongst the best known polysaccharides.

Cellulose

Cellulose, $(C_6H_{10}O_5)_n$, is a natural polymer of β - d(+) glucose, most widely distributed in plants where it is found in wood, cereal, straw up to about 50% in association with lignin, hemicellulose (polysaccharide of other sugars) and non-cellulose substances. Cotton which contains almost 95% of cellulose, is the oldest natural fiber and is in common use till today. The fiber is resistant to alkali and can be easily dyed. Filter paper is nearly pure cellulose.

Plants synthesize cellulose out of one of the most basic of all organic substances-glucose, a simple sugar (monosaccharide) which is found throughout the vegetable kingdom and even in the blood of animals. Nature converts glucose into cellulose (in plants) when thousands of glucose molecules are linked together through hydroxyl groups at C_1 and C_4 in a long chain forming a polyanhydride (giant cellulose molecule) and several water molecules (by a process similar to polycondensation). The following is the structure of the repeat unit in cellulose:

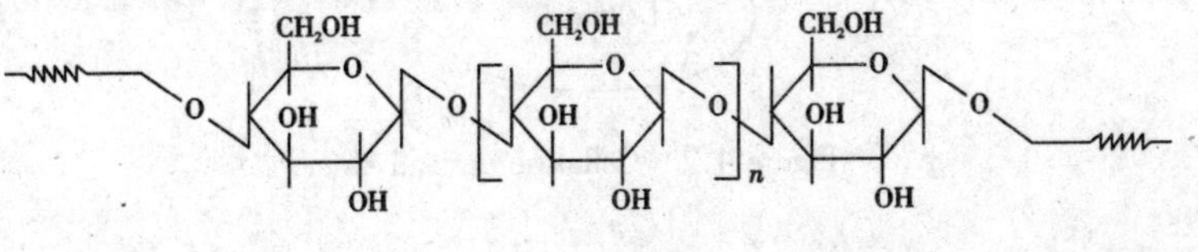

Despite containing several hydroxyl groups (three in each monomeric unit), cellulose is unexpectedly water insoluble. The reason for this is the stiffness for the chains and hydrogen bonding between two —OH groups on adjacent chains. Cellulose differs from starch (polymer of α-glucose) only in arrangement of hydroxyl group at carbon number one in the monomer glucose unit. Both are the polymers of glucose. The two polysaccharides in spite of having only slight difference, possess enormously different properties. Starch being the chief constituent of diet, gets easily hydrolyzed by body enzymes to glucose which finally yields CO_2 and H_2O atoms with the energy, the later being used for sustaining body functions. Cellulose nevertheless is of considerable human digestive value in that it adds "bulk" to the intestinal contents, thereby stimulating peristalsis and eliminates food residues. It thus provides indigestible but necessary roughage in the human diet, and the nourishment for such ruminant animals as sheep and cows. Indirectly, therefore it also provides nourishment for humans. Sheep and cows digest the cellulose and absorb it into their body tissues and thus when humans consume meat, cellulose is used up by them.

It has the property of forming long threads and for this reason is largely incorporated as raw material in textile industries. Cellulose is highly stereo specific macromolecule. Considering the geometry of the polymer molecule, cellulose can be treated as non-planar molecule, but with a screw-axis and with each glucose unit being at right angles to the previous one. The free rotation about the C—O—C link does not occur. This and the close packing of the atoms result in a fully extended form of conformation. Moreover large intra-and inter-molecular forces (as a result of formation of hydrogen bonds) also favor the close packing and thus cellulose possesses an unusually high degree of crystallinity. The melting temperature of crystalline cellulose is far above its decomposition temperature. The fiber forming properties of cellulose can be accounted by its unique conformation. The fully extended cellulose chains are further bonded together to give rise to a sheet structure. A huge amount of cellulose is used in the paper manufacture. Besides its applications in textile and paper industries, cellulose finds use as celluloid (a hard plastic), cellophane (film) for various purposes. These materials are called semi-synthetic, because they are obtained from the raw cellulose. The synthesis of various cellulose derivatives by chemical reactions is known as regeneration. The various regenerated forms of cellulose are shown in Figure 4.2.

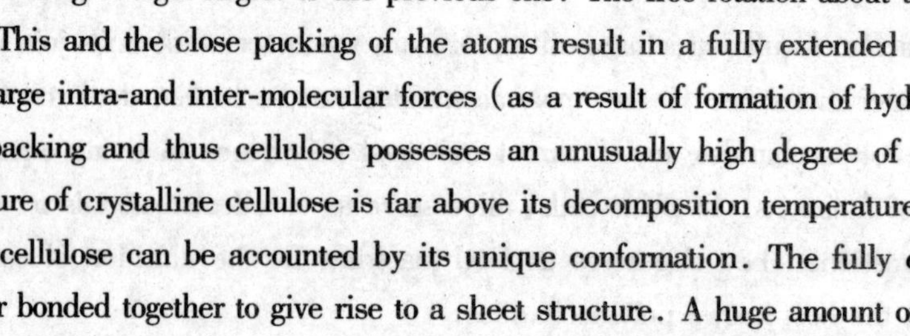

Figure 4.2 Cellulose derivatives

Chitin and chitosan

Chitin, a naturally abundant mucopolysaccharide, and the supporting material of crustaceans, insects, etc., is well known to consist of 2-acetamido-2-deoxy-β-D-glucose through a C_1-C_4 ($1 \rightarrow 4$) linkage. Chitin can be degraded by chitinase. Its immunogenicity is exceptionally low, in spite of the presence of nitrogen. It is a highly insoluble material resembling cellulose in its solubility and low chemical reactivity. It may be regarded as cellulose with hydroxyl at position C_2 replaced by an acetamido group.

Like cellulose, it functions naturally as a structural polysaccharide. Chitin is a white, hard, inelastic, nitrogenous polysaccharide and the major source of surface pollution in coastal areas. Chitosan is the *N*-deacetylated derivative of chitin, although this *N*-deacetylation is almost never complete. A sharp nomenclature with respect to the degree of *N*-deacetylation has not been defined between chitin and chitosan. The structures of chitin and chitosan are shown in Figure 4.3.

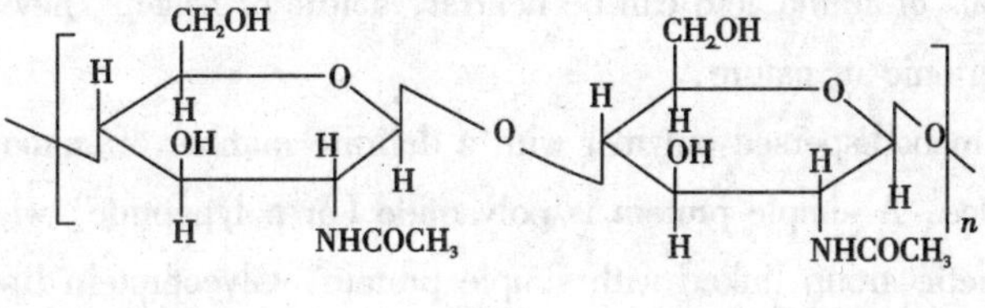

chitin

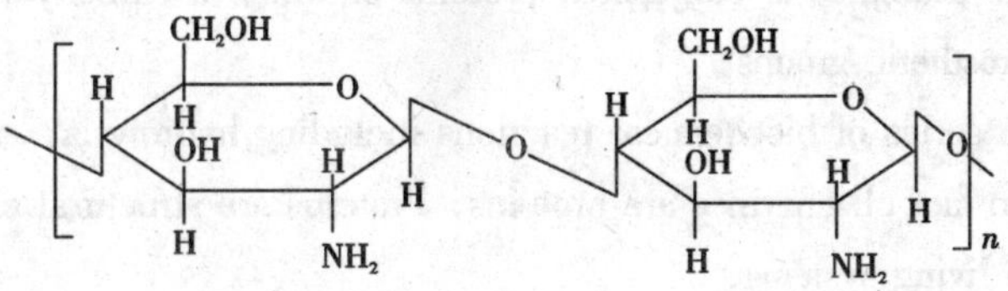

chitosan

Figure 4.3 Structures of chitin and chitosan

Chitin and chitosan are of commercial interest due to their high percentage of nitrogen (6.89%) compared to synthetically substituted cellulose (1.25%). This makes chitin a useful chelating agent. As most of the present-day polymers are synthetic materials, their biocompatibility and biodegradability are much more limited than those of natural polymers such as cellulose, chitin, chitosan and their derivatives. However, these naturally abundant materials also exhibit a limitation in their reactivity and processability. In this respect, chitin and chitosan are recommended as suitable functional materials, because these natural polymers have excellent properties such as biocompatibility, biodegradability, non-toxicity, adsorption properties, etc. Recently, much attention has been paid to chitosan as a potential polysaccharide resource. Although several efforts have been reported to prepare functional derivatives of chitosan by chemical modifications, very few attained solubility in general organic solvents and some binary solvent systems. Chemically modified chitin and chitosan structures resulting in improved solubility in general organic solvents have been reported by many workers.

Chitin and chitosan have a wide range of applications. They may be employed, for example, to solve numerous problems in environmental and biomedical engineering. Chitin derivatives including partially deacetylated chitosan can be easily molded to various forms and their derivatives are digested *in vivo* by lysozomal enzymes. Thus, it appears that this material can be a most interesting candidate for use as a carrier of a variety of drugs for controlled release applications. Lately, the transdermal absorption promoting characteristics of chitosan have been exploited, especially for nasal and oral delivery of polar drugs to include peptides and proteins and for vaccine delivery. These properties, together with the very safe toxicity profile, make chitosan an exciting and promising excipient for the pharmaceutical industry for present and future applications.

4.2.2 Proteins

Proteins are the natural polymers made of α-amino acid units through peptide (—CONH—) linkage. In total about twenty of these amino acids occur in nature (of which eight are essential to sustain life). Except the simplest amino acid glycine, all other amino acids are optically active and are of L-series. Aqueous solutions of amino acid can be neutral, acidic or basic. They also form dipolar structure and thus are zwitterionic in nature.

Each protein is a monodispersed polymer with a definite number of ammo acids linked with each other in distinct sequence. A simple protein is polyamide (or polypeptide) whereas a conjugated protein consists of a prosthetic group linked with simple protein. Glycoprotein lipoprotein, nucleoprotein and haemoglobin are the examples of conjugated proteins of which a carbohydrate, a lipid, a nucleic acid or haem are the prosthetic groups.

Proteins catalyze a series of biochemical reactions including hydrolysis, group transfer, oxidation and biosynthesis etc. In fact all enzymes are proteins. Proteins are structural elements and function as protective agents in the living systems.

Proteins may be inter-molecularly hydrogen bonded (fibrillar proteins) e.g. keratin of the hair and nails, collagen of the connective tissues, elastin of the sinuses and arteries, fibroin of silk thread; myosin, the muscle protein or intra-molecularly hydrogen bonded (globular proteins) e.g. hormones like insulin and glucagon, enzymes like trypsin, invertase and papain etc., respiratory proteins such haemoglobin, myoglobin, aibumins e.g. egg albumin, bovine serum albumin and antibodies. Fibrillar proteins are water insoluble whereas globular proteins are usually water soluble.

All proteins are hydrolyzed finally to a mixture of α-amino acid from which these were built. The reagent ninhydrin gives characteristic coloration to amino acids and thus may be determined chromatographically (a method developed by Nobel laureates Martin and Synge).

The structure of proteins is very complex. Primary structure of proteins shows the chemical formulae of the amino acids arranged in a definite sequence through peptide linkage. The sequence of amino acids in the protein molecule was first successfully examined by Sanger who was awarded two Nobel prizes in chemistry. Secondary structure of proteins describes the molecular shape (or conformation) of a protein molecule. Nobel laureate Linus Pauling, has demonstrated that a right handed intra-molecularly hydrogen bonded helical arrangement (α-helix) is an important secondary structure of proteins when many bulky pendant groups are present on the main chain. A β-arrangement or pleated sheet conformation predominates when small pendant groups are present on the chain. The α- and β-forms of proteins enzyme are shown in Figure 4.4.

Tertiary structure designates the shape of folding which results due to the sulfur-sulfur crosslinks between macromolecular chains. The specificity of enzyme action in fact depends on tertiary structure. Proteins lose their physiological activity because of the disruption of folded structure (denaturation of proteins). Individual protein chain agglomerates into supramolecular system and this mode of agglomeration is called "quaternary structure".

Although polypeptides can be synthesized by simply heating α-amino acids but the products obtained are random mixtures unless a single amino acid is used. A solid phase technique developed by

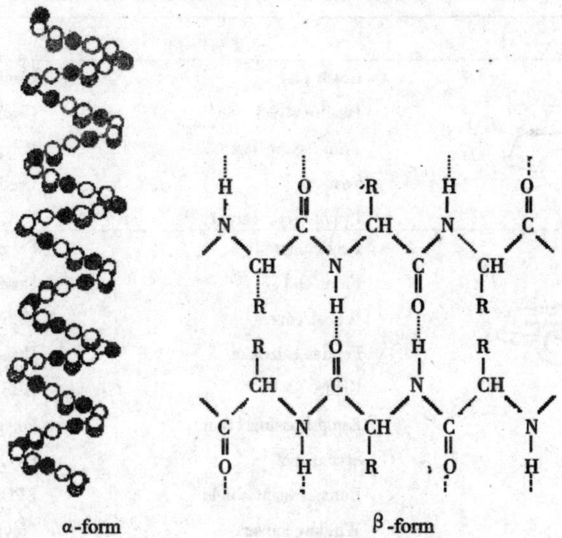

Figure 4.4 The α- and β-form of proteins

Merrifield (Nobel laureate, 1984) is a novel method to synthesize polypeptides with desired ammo acid sequence. In Merrifield synthesis, all reactions take place on the surface of crosslinked polystyrene beads and the entire reaction for synthesizing polypeptides with many programmed sequences of amino acids can be carried out automatically in a simple vessel with a possibility of neat isolation of any intermediate products.

4.3 Thermoplastics

Each year the United States consumes about 79,000 million pounds of plastic and synthetic resins, or about 260 pounds for every citizen. The use of lightweight plastics has helped increase gas mileage in automobiles (Table 4.5), and this trend will increase with plastic car bodies being more widely used in the near future.

It is important to recognize that of the total yearly U.S. oil and gas consumption, 60% is used as stationery fuels in home heating and fuel to run power plants, 33% is used as transportation fuels, and 7% is used for the manufacture of petrochemicals, including fertilizer, rubber, paints, fibers, solvents, and medicines. Only about 2.5% is employed for polymer applications, yet this 2.5% makes possible the production of many useful products.

All polymers are classified as either thermoplastics, that is, linear or branched polymers that can be reversibly softened by heating and solidified by cooling, or thermosets, that is, crosslinked polymers that can not be softened by heating without degradation. The word plastics derived from the Greek word plastikos, meaning able to be molded. Both thermoplastics and thermoset prepolymers can be molded into desirable shapes.

Table 4.5 Major plastics applications in automobiles

Area	Applications	Materials (Usual)
Interior	Crash pad	Urethane, ABS, PVC
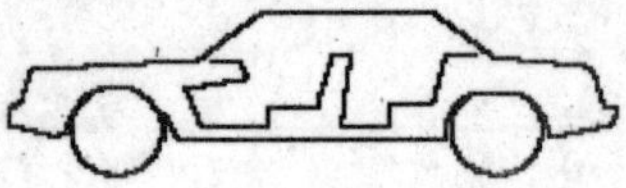	Headrest pad	Urethane, PVC
	Trim, glove box	Polypropylene, PVC, ABS
	Seat	Urethane
	Upholstery, carpet	PVC, nylon
Exterior	Fender apron	Polypropylene
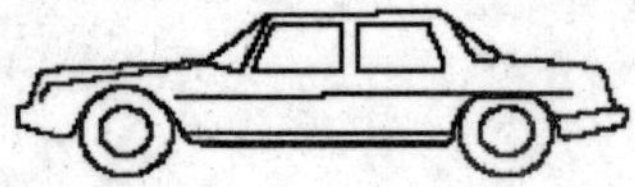	Front end	Unsaturated polyester
	Wheel covers	ABS, poly(phenylene oxide)
	Fender extension	Unsaturated polyester, nylon
	Grille	ABS, poly(phenylene oxide)
	Lamp housing (rear)	Polypropylene
	Styled roof	PVC
	Bumper sight shield	EPDM rubber, urethane
	Window louvers	Poly(butylene terephthalate)
Under the hood	Ducts	Polypropylene
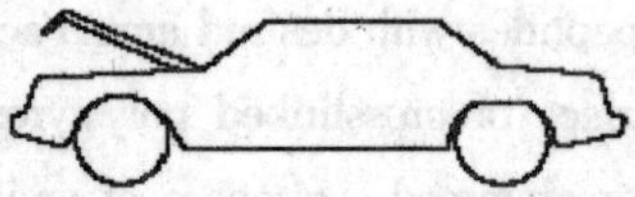	Battery case	Polypropylene
	Fan shroud	Polypropylene
	Heater and air conditioning	Unsaturated polyester
	Electrical housing and wiring	Phenolic, PVC, silicone
	Electronic ignition components	Poly(butylene terephthalate)

The difference between thermoplastics and engineering plastics is often a thin line and depends on whether the material is to be cut, drilled, and so on. Some plastics like Teflon™ can be drilled and cut but with some difficulty so that it is between a thermoplastic and an engineering plastic. Here Teflon will be covered as a thermoplastic. Both thermoplastics and engineering plastics are thermoplastics. Those vinyl polymers that do not contain highly polar (like carbonyls) and hydrogen-bonding units (like amides) have relatively weaker secondary forces between chains and are thus included in this section.

The production of thermoplastics and engineering plastics for the year 2000 is given in Table 4.6. Most of the thermoplastics are derived from vinyl reactants. The formation of the vinyl polymers occurs whereby the (bond of the carbon-carbon double bond, C=C, is broken and new σ bonds are formed that link the monomer units together, forming the polymer chain. The formation of the additional σ bonds creates energy. Thus, these vinyl polymer forming reactions are exothermic; that is, they create energy in the form of heat. Large-scale vinyl reactions need to be cooled because of the formation of this heat.

Vinyl polymerizations are started, initiated, by something that breaks the π bond like heat or high energy light. In the dentist's office an ultraviolet lamp is often used to break this double bond, resulting in the dental cement or filling to be cured or crosslinked through breakage of the π bonds and formation of new σ bonds that crosslink or set the dental material. For many reactions, a compound is added to the monomer mixture that has chemical bonds that are easily broken to give products that start the reaction process. These compounds are called initiators because they initiate or start the polymer forming reaction.

Table 4.6 U.S. production of thermoplastics and engineering plastics, 2000

Plastics	Production (Millions of Pounds)
Acrylonitrile-butadiene-styrene, ABS	3,100
Polyamides, nylons	1,400
Polyesters	4,400
Polyethylene, highdensity	15,400
Polyethylene, lowdensity	17,900
Polystyrene	6,600
Styrene-acrylonitrile	124
Polypropylene	15,400
Poly(vinylchloride) and copolymers	14,300

By comparison, engineering thermoplastics are generally formed through condensation reactions that require heating with the overall reaction begin endothermic or heat consuming.

4.3.1 Polyethylene

Polyethylene is probably the polymer you see most in daily life. Polyethylene is the most popular plastic in the world. This is the polymer that makes grocery bags, shampoo bottles, children's toys, and even bullet proof vests. For such a versatile material, it has a very simple structure, the simplest of all commercial polymers. A molecule of polyethylene is nothing more than a long chain of carbon atoms, with two hydrogen atoms attached to each carbon atom as shown in Figure 4.1.

Sometimes it's a little more complicated. Some of the carbons, instead of having hydrogens attached to them, will have long chains of polyethylene attached to them. This is called branched, or low-density polyethylene (LDPE) (Figure 4.5(a)). When there is no branching, it is called linear polyethylene, or high-density polyethylene (HDPE) (Figure 4.5(b)). Linear polyethylene is much stronger than branched polyethylene, but branched polyethylene is cheaper and easier to make.

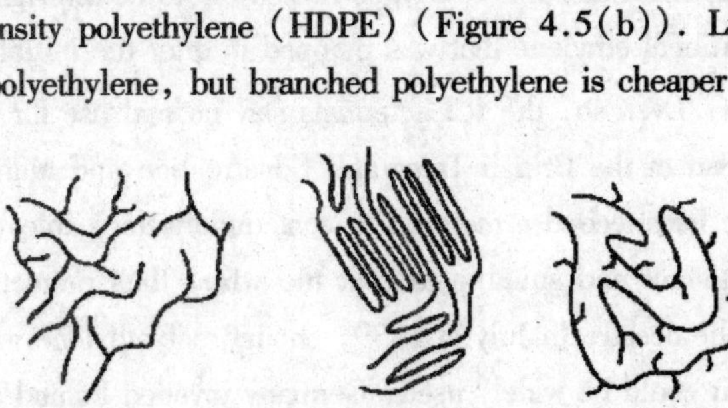

Figure 4.5 Representations of different forms of polyethylene to illustrate branching and nonbranching aspects. (a) Low-density polyethylene (LDPE). (b) High-density polyethylene (HDPE). (c) Linear low-density polyethylene copolymer with 1-butene (LLDPE). The 1-butene discourages crystallization

Linear polyethylene is normally produced with molecular weights in the range of 200,000 to 500,000, but it can be made even higher. Polyethylene with molecular weights of three to six million is referred to as ultra-high molecular weight polyethylene (UHMWPE). UHMWPE can be used to make fibers which are so strong they replaced Kevlar for use in bullet proof vests. Large sheets of it can be used instead of ice for skating rinks.

Polyethylene is vinyl polymer, made from the monomer ethylene. Branched polyethylene is often made by free radical polymerization. Linear polyethylene is made by a more complicated procedure called Ziegler-Natta polymerization. UHMWPE is made using metallocene catalysis polymerization. But Ziegler-Natta polymerization can be used to make LDPE, too. By copolymerizing ethylene monomer with an alkyl-branched comonomer, one gets a copolymer which has short hydrocarbon branches. Copolymers like this are called linear low-density polyethylene (LLDPE) (Figure 4.5(c)).

History

Polyethylene for commercial attention was probably initially synthesized by M. E. P. Friedrich while he was a graduate student working for Carl S. Marvel in 1930 when it was an unwanted by product from the reaction of ethylene and a lithium alkyl compound. In 1932, British scientists at the Imperial Chemical Industries (ICI) accidently made polyethylene while they were looking for products that could be produced from the high-pressure reaction of ethylene with various compounds. In March 1933, they found the formation of a white solid when they combined ethylene and benzaldehyde under high pressure. They correctly identified the solid as polyethylene. They attempted the reaction again, but with ethylene alone. Instead of again getting the waxy white solid, they got a violent reaction and the decomposition of the ethylene. They delayed their work until December 1935 when they had better high-pressure equipment. At 350 °F, the pressure inside of the reaction vessel containing the ethylene decreased, consistent with the formation of a solid. Because they wanted to retain the high pressure, they pumped in more ethylene. The observed pressure drop could not be totally due to the formation of polyethylene, but something else was contributing to the pressure loss. Eventually they found that the pressure loss was also due to the presence of a small leak that allowed small amounts of oxygen to enter into the reaction vessel. The small amounts of oxygen turned out to be the right amount needed to catalyze the reaction of the additional ethylene that was pumped in after the initial pressure loss — another major discovery by accident. Even so, the ICI scientists saw no real use for the new material.

By chance, J. N. Dean of the British Telegraph Construction and Maintenance Company heard about the new polymer. He had needed a material to coat underwater cables. He reasoned that polyethylene would be water-resistant and suitable to coat the wire, thus protecting it from the corrosion caused by the saltwater in the ocean. In July of 1939, enough polyethylene was made to coat one nautical mile of cable. Before it could be widely used, Germany invaded Poland and polyethylene production was diverted to making flexible high-frequency insulated cable for ground and airborne radar equipment. Polyethylene was produced, at this time, by ICI in the United Kingdom and by DuPont and Union Carbide for the United States.

Polyethylene did not receive much commercial use until after the war when it was used in the manufacture of film and molded objects. Polyethylene film displaced cellophane in many applications being used for packaging products, textiles, frozen and perishable foods, and so on. This polyethylene was branched and had a relatively low softening temperature (below 212 °F), preventing its use for materials where boiling water was needed for sterilization.

Karl Ziegler, director of the Max Planck Institute for Coal Research in Muelheim, Germany was extending early work on polyethylene attempting to get ethylene to form polyethylene at lower pressures and temperatures. They found compounds that allowed the formation of polyethylene under much lower pressures and temperatures. Furthermore, these compounds produced a polyethylene that had fewer branches and a higher softening temperature (above 212 °F), allowing the material to be sterilized when needed.

The branched polyethylene is called low-density, high-pressure polyethylene because of the high pressures usually employed for its production; and because of the presence of the branches, the chains are not able to closely pack, leaving voids and subsequently producing a material that has a lower density in comparison to low-branched polyethylene.

Giulio Natta, a consultant for the Montecatini company of Milan, Italy applied the Zeigler catalysts to other vinyl monomers such as propylene and found that the polymers had higher densities, exhibited higher melting points, and were more linear than those produced by the classical techniques such as free radical initiated polymerization.

Ziegler and Natta shared the Nobel prize in 1963 for their efforts in the production of vinyl polymers using what we know today as solid-state stereoregulating catalysts.

Today there exist a number of polyethylenes that differ mainly in the amount and length of branching as well as in chain length and chain length variation. Properties of typcal polyethylenes are shown in Table 4.7.

Table 4.7 Properties of typical polyethylenes

Properties	HDPE	LDPE	LLDPE	PP
Melting point (T_m, °F)	275	200	252	338
Glass transition temp. (T_g, °F)	—	20	—	—
Processing temp. (°F)	450	400	450	450
Molding pressure (10^3 psi)	15	10	10	15
Maximum resistance to continuous heat (°F)	175	110	130	125
Tensile strength (10^3 psi)	35	30	33	50
Flexural strength (10^3 psi)	30	—	—	60
Tensile modulus (10^3 psi)	155	35	45	50
Elongation(%)	200	300	400	400
Flexural modulus (10^3 psi)	150	30	50	200
Specific gravity	0.96	0.91	0.93	0.91

[a]psi/0.145 = kPa (kilopascals).

High-density polyethylene

The alkanes, such as paraffin wax, high-density polyethylene (HDPE), and other polyolefins, have the empirical formula, $H\text{-}(CH_2)_n\text{-}H$. Nevertheless, the degree of polymerization of paraffin wax and lower-molecular-weight alkanes is too low to permit entanglement of the polymer chains. Another highly branched polyolefin, called elasterite, occurs naturally in the fossil Fungus subterraneus but was never used as a commercial plastic.

The first commercial HDPE was produced independently by J. Hogan and R. Banks in the United

States and later by Ziegler in Germany in the early 1950s. Ziegler's synthesis was related to that used by Marvel. K. Ziegler, and co-workers used aluminum triethyl [$Al(C_2H_5)_3$] and titanium trichloride ($TiCl_3$) for their polymerization catalyst in what they called the building-up reaction. HDPE is now a commercial plastic, with over 7.7 million tons being produced annually in the United States.

Density is related to how tightly material can be packed. Thus linear polyethylene (HDPE) can be tightly packed since linear chains can be efficiently folded as noted in Figure 4.5(b). Conversely, branched polyethylene (LDPE) packs less firmly as a result of the presence of the branches, which prohibit close, regular folding. Thus, more ethylene units can be packed within a specific volume for linear polyethylene than for branched polyethylene, resulting in a higher weight per volume and consequently higher density for linear polyethylene.

Specific gravity is simply the ratio of the density of the material compared to the density of water. Employing the units of g/cc, the density and specific gravity of a material are essentially the same since the density of water is 1.0 g/cc and a number divided by one is the number itself.

We can get a better ideal of what linear polyethylene looks like on a molecular level in Figure 4.6. It is usually crystalline, forming a close-packed molecular package. This package is strong, brittle and relative to amorphous structures. It does not allow molecules to pass through. Furthermore, this tightly packed molecular bundle is not easily penetrated by unwanted chemicals such as acids and bases. Also, because these acids and bases, like acetic acid in vinegar and citric acid from lemons, are water-loving (hydrophillic) and the ethylene chain is nonpolar and does not love water (hydrophobic), these acids and bases are not attracted. Thus, polyethylene is useful in our homes, where small amounts of acids and bases are a part of life. By comparison, oils like motor oils are themselves composed of hydrocarbon chains so that they are polyethylene-like, but because of the close packing the linear crystalline polyethylene does not allow the motor oil to pass through or to do substantial damage.

While linear polyethylene has a great tendency to form the closely packed crystalline package, it can be made to take an amorphous structure through mixing the melted polymer and then quick-cooling the material locking in the amorphous structure (Figure 4.7).

The amorphous chain in Figure 4.7 has about 50% empty space. Even so, it is largely impervious to most chemicals such as oxygen. The thickness of a page of this book is about 0.05 mm. A sheet of polyethylene with the same thickness (0.05 mm) would have about 10^{11} layers of methylene units. Thus, while it looks like there are large holes, the sum total presents a large maze for the oxygen molecules to go through. Even so, some oxygen gets through. Crystalline polyethylene has many fewer empty spaces and allows less oxygen to get through.

The thickness of an ordinary envelope is also on the order of 0.05 mm in thickness. Yet the anthrax bacteria, a much larger array of atoms than the simple oxygen, is able to get through the pores of the envelope and contaminate the machinery and personal that comes in contact with the anthrax-containing envelope. The difference is that the polyethylene film is a continuous layer upon layer of polymer chains while the envelopeis composed of paper fibers which, even when matted together, have, on a molecular level, holes large enough to allow the anthrax to get through. The matted paper fibers are not continuous on a molecular level and even though it appears to the naked eye to be continuous, it is

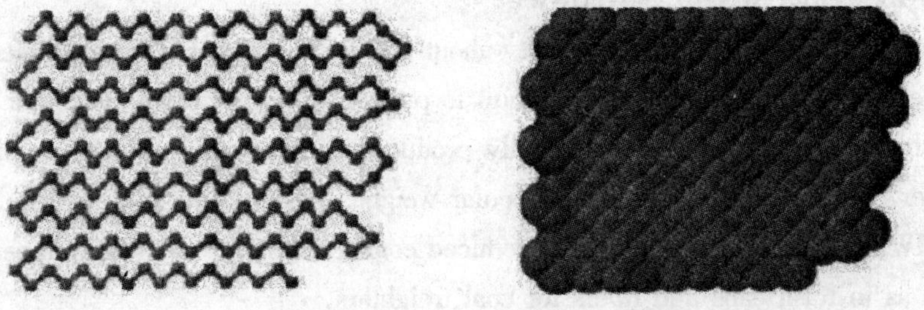

Figure 4.6 Two-hundred-carbon crystalline linear polyethylene chain

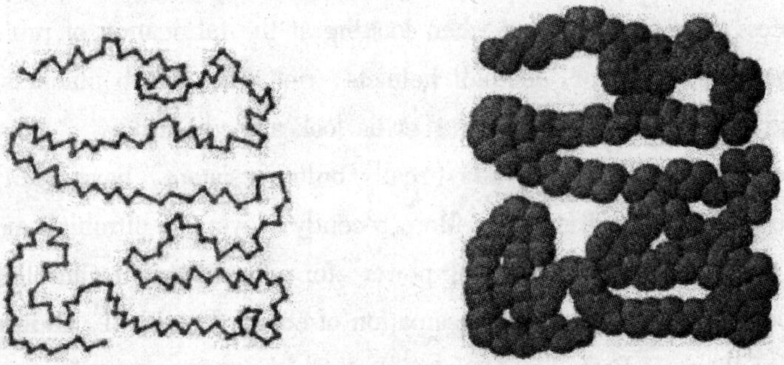

Figure 4.7 Two-hundred-carbon polyethylene chain in an amorphous orientation

not.

Thus, both crystalline and amorphous polyethylenes are good barriers to most materials.

HDPE is produced using catalysts and has normally less than 15 (usually 1 to 6) short branches per 1,000 ethylene units. It has a melting point of about 130 °C, so it can be made into objects that need to be sterilized through the use of boiling water. Typical products are bottles, trays, drums, tanks, cans, pails, house wares, toys, food containers, conduit, wire and cable coating, foam, insulation for coaxial and communication cables, pipes, bags, films, and crates.

Low-density polyethylene

As shown by the data in Table 4.7, LDPE has a lower modulus (is more flexible) and has a lower melting point than HDPE. It is also less stable than HDPE, though still quite stable in comparison to most other giant molecules, to acids, bases, oils, and so on; and gases such as carbon dioxide and oxygen are able to penetrate sheets and films of LDPE more readily because these invading molecules have a greater opportunity to penetrate the less tightly packed structure (Figure 4.5(c)). LDPE has between 50 and 150 short alkyl branches for every 1,000 ethylene units. This branching is sufficient to discourage crystalline formation, resulting in a material that is about 50% amorphous.

LDPE films are nearly clear even though they contain a mixture of crystalline and amorphous regions. This is because the crystalline portions are space-filling and not isolated spherulites, allowing a largely homogeneous structure with respect to refractive index that results in the material being transparent. In fact, the major reason why LDPE films appear hazy or not completely transparent is because of the roughness of the surface and is not due to the light scattering of the interior material.

LDPE is used for bags like the ones you get at the checkout counters in stores; it is also used for packaging products, films, sheeting, piping, industrial containers, and household items.

Ultrahigh-molecular-weight polyethylene

A minimum or threshold molecular weight (about 100 DP) is required for entanglement of HDPE. Since high-molecular-weight polymers are difficult to process, polymers with molecular weights slightly above the threshold molecular weight are usually produced commercially. However, the toughness of HDPE and other polymers increases with molecular weight. Hence, ultrahigh-molecular-weight polyethylene (UHMWPE) (DP = 1 million) is produced commercially for use where unusual toughness is essential, such as in trash cans and liners for coal freighters.

Polymers, because of their "connectiveness," are able to readily share energy and distortions, thereby redistributing the energy and distortions throughout the chain and between chains. This acts to dissipate impact forces. This is illustrated when looking at the fabrication of protective armor such as vests, vehicle exteriors, body armor, football helmets, riot gear, bomb blankets, explosive containment boxes, bus and taxi shields, and so on. Let us look at body armor.

Many of the so-called bullet-proof vests (really bullet-resistant) have been made of very tough and strong polycarbonates such as Kevlar™. More recently, layers of ultrahigh-molecular-weight polyethylene were found to a have similar "stopping power" for a lesser weight, and today many of the vests are made using polyethylene sheets. Rapid dissipation of energy is critical, allowing the impact energy to be spread over a wide area. Body armor material should be strong enough so as not to immediately break when the bullet hits but have enough contact with other parts of the body armor to allow ready transfer of some of the impact energy. If the material can adsorb some of the energy though bond breakage or heating, then additional energy can be absorbed at the impact site. Along with high strength, the material should have some ability to stretch, allowing the material to transfer some of the energy to surrounding material. If connective forces between components are too strong, total energy dissipation is reduced because strong bonding between the various parts of the body armor discourages another form of energy sharing, that of having the parts slide past one another, thereby redistributing the impact energy. Thus, a balance is needed between material strength, strength holding the various components of the body together, and the ability to readily dissipate the impact energy. Certain sequences of layered material have been found to be more effective at dissipation of the impact energy than others. One of the employed combinations contains sheets of strong and more rigid aramid fiber adjacent to sheets of ultrahigh-molecular-weight polyethylene that is less strong but flexible and stretchable.

Another factor is the breakup of the projectile. Again, super strong giant molecules composed of composites containing boron carbide ceramics and aramids, ultrahigh-molecular-weight polyethylene, or fibrous glass are effective at breaking up the projectile.

Linear low-density polyethylene

Commercial copolymers in which both ethylene and 1-butene ($H_2C{=}CHCH_2CH_3$) are present as repeating units in the polymer chain are linear (Fig. 4.5), but because of the bulky pendant (C_2H_5) groups they occupy greater volume and have a lower specific gravity than HDPE. Linear low-density polyethylene (LLDPE) may be produced at low pressure in the gaseous phase or in solution. Higher homologues such as 1-octene [$H_2C{=}CH(CH_2)_5CH_3$] may also be used as the comonomers in LLDPE. New coordination catalysts, which are related to those used for making HDPE, are also used in the production of LLDPE.

LDPE is characterized by good flexibility and hence can be used as a film and squeeze bottles. HDPE is stiffer and more heat-resistant and is used as rigid pipe. LLDPE is stronger than LDPE and can be used as thinner films for making bags, for example.

Plastomer is the name given to copolymers of ethylene that have a little crystallinity but are largely amorphous. They are also called very low density polyethylene (VLDPE). They are more elastic than LLDPE but less stiff. They are used as a sealing layer in film applications and controlled permeation packaging for vegetables and fruits.

Other copolymers of ethylene

In addition to copolymers of 1-olefins, such as LLDPE, there are several other commercial copolymers of ethylene. The copolymer of ethylene and vinyl acetate is an amorphous copolymer that may be cast as a clear film or used as a melt coating. The copolymer of ethylene and methacrylic acid [CH_2 = $C(CH_3)COOH$] is also a moldable thermoplastic. This copolymer, when partially neutralized to form monovalent and divalent metal-containing materials, is called an ionomer. These ionomer salts have a stable crosslinked structure at ordinary temperatures but can be injection-molded. These tough copolymers are used as golf ball covers in place of balata.

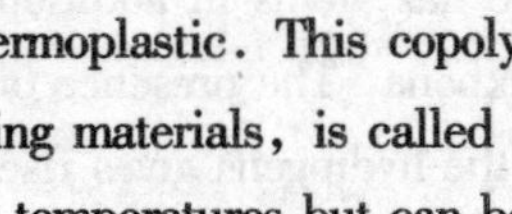

ethylene-methacrylic acid copolymers (ionomers)

Both HDPE and polypropylene are high-melting crystalline polymers. However, the random copolymer of these two comonomers is an amorphous, low-melting elastomer. It is customary to add a crosslinking monomer, such as dicyclopenta-diene, to the comonomers to produce a vulcanizable elastomer (EPDM). EPDM is used as the white sidewalls of tires and as single-ply roofing material.

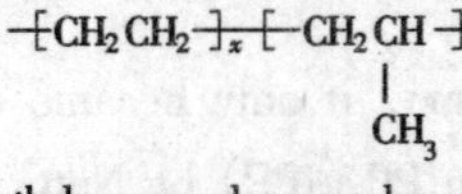

ethylene-propylene copolymer

These ethylene-propylene copolymers are also employed in other automotive applications such as radiator and heater hoses, seals, mats, weather strips, bumpers, and body parts. Nonautomotive applications include coated fabrics, gaskets and seals, hoses and wire, and cable insulators.

The block copolymer of ethylene and propylene, which contains long sequences of ethylene and propylene repeating units, is a clear, moldable copolymer and is used in place of HDPE in many applications. Its specific gravity is similar to that of LDPE.

4.3.2 Polypropylene

Nobel laureate K. Ziegler patented HDPE but failed to include polypropylene (PP) in his patent application. However, many other chemists used the Ziegler catalyst [$TiCl_3 \cdot (Al(C_2H_5)_3$] to produce PP in the early 1950s. Nobel laureate G. Natta of Montedison, W. Baxter of DuPont, and E. Vanderburg of Hercules filed for patents for the production of PP using the Ziegler catalyst. J. Hogan and R. Banks of Phillips and A. Zletz of Amoco filed for patents using supported metal oxide catalysts. In 1973 the U.S. Patent Office granted a patent for PP to Natta, but reversed its decision in favor of Hoganand Banks in 1983.

$$-\!\!\left[CH_2\underset{\underset{CH_3}{|}}{CH} \right]\!\!-_n$$

polypropylene (PP)

Polypropylene (PP) is one of the three most heavily produced synthetic polymers. This abundance of PP is called for because of its variety and versatility being employed today in such diverse applications as a film in disposable diapers and hospital gowns to geotextile liners; plastic applications as disposable food containers and automotive components; and fiber applications such as carpets, furniture fabrics, and twine. Typical properties are given in Table 4.7.

Unlike polyethylene, polypropylene has atoms in addition to hydrogen attached to the polymer backbone. The presence of the methyl group substituting for one of the hydrogens gives rise to several different structural isomers or "tacticities," called (a) isotactic, and (b) syndiotactic isomers, which are both regular or ordered structure, and (c) atactic isomers, which have only a somewhat random arrangement of methyl groups (Figure 4.8). The regular structures, isotactic and syntiotactic, allow the polymer chains to more readily come closer together, forming crystalline structures. These crystalline structures are reflected in so-called stereoregular PP being stronger, less permeable to acids, oils, and gas molecules, and higher-melting. The disordered random atactic, (where the prefix "a" means having nothing to do with) structure gives a less crystalline product.

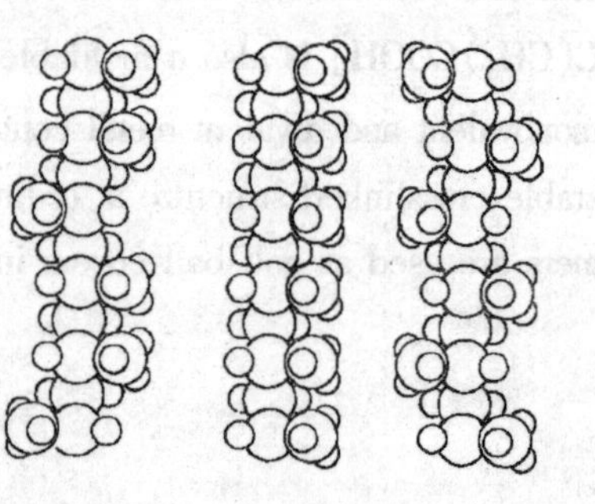

Figure 4.8 Atactic (left), isotactic (middle), and syndiotactic (right) polypropylene

While PP was produced for sometime, it only became commercially available in the late 1950s with the production of somewhat isotactic PP (iPP) by Natta and co-workers at Phillips. The first PP was not highly crystalline because the tacticity was only approximate; but with the use of the Natta-Zeigler catalysts, iPP was produced that contained greater amounts of stereoregular material with a corresponding increase in crystallinity and associated properties such as an increased stiffness, better clarity, and a higher distortion temperature. Today, with better catalysts, including the soluble metallocene catalysts, the tacticity has been increased so that 99% isotactic material can be readily produced. The more traditional Zeigler-Natta catalysts systems today have activities as high as producing 100 000 pounds iPP per pound of catalysts eliminating the need for catalyst removal.

While most of the PP used today is of the regular or stereoregular form, atactic or amorphous forms of PP are also used. Initially, atactic PP (aPP) was obtained as a byproduct of the production of iPP. As an inexpensive byproduct it is used as a modifier for asphalt for roofing and in adhesives. As the effectiveness of catalyst systems become better, less aPP is available so that today some aPP is intentionally made for these applications.

Because iPP is stiff and has a relatively high T_g (about 0 °C), some effort has involved lowering of the T_g and to achieve greater toughness. One approach is to employ a little ethylene in the polymerizing mixture, thus creating a copolymer that is largely iPP but with enough polyethylene (PE) to effectively lower the amount of crystallinity. Such copolymers are called reactor copolymers (RCPs). Another approach is to blend iPP with rubber. These materials form the important class of ethylene-propylene copolymers.

The use of metallocene catalyst systems allows the formation of an important group of copolymers from the use of α-olefins as comonomers with propylene. These catalysts also allow the production of wholly PP block copolymer elastomers that contain blocks of aPP and iPP. The aPP blocks act as the soft and/or non-crosslinked portion, while the iPP blocks act as the hard and/or physically crosslinked portion.

4.3.3 Polystyrene

Styrene monomer was discovered by Newman in 1786. The initial formation of polystyrene (PS) was by Simon in 1839. While PS was formed almost 175 years ago, the mechanism of formation was not discovered until the early twentieth century. Staudinger, using styrene as the principle model, identified the general free radical polymerization process in 1920. Initially, commercialization of PS, as in many cases, awaited the ready availability of the monomer. While there was available ethyl benzene, it underwent thermal cracking rather than dehydrogenation until the appropriate conditions and catalysts were discovered.

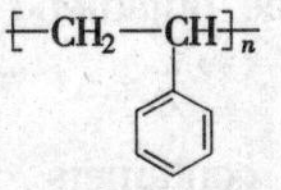

Dow first successfully commercialized polystyrene formation in 1938. While most commercial PS has only a low degree of stereoregularity, it is rigid and brittle because of the resistance of easy movement of the more bulky phenyl-containing units in comparison, for example, to the methyl-containing units of polypropylene (Figure 4.9). This is reflected in a relatively high T_g of about 212 °F for polystyrene. It is transparent because of the low degree of crystalline formation.

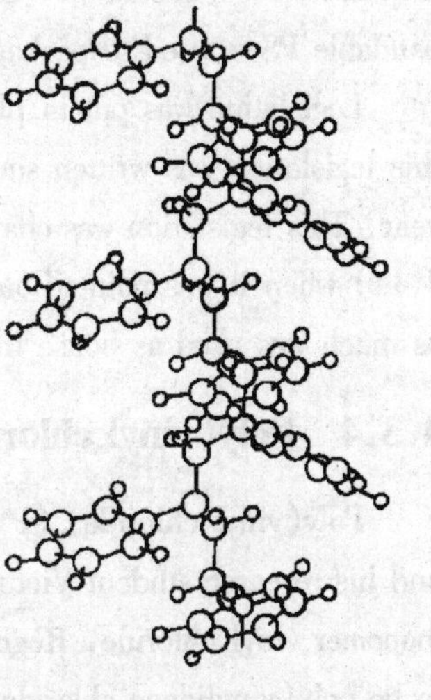

Figure 4.9 Ball-and-stick model of polystyrene

While PS is largely commercially produced using free radical polymerization, it can be produced by all four of the major techniques: anionic, cationic, free radical, and coordination-type systems. All of the tactic forms can be formed employing these systems. The most important of the tactic forms is syndiotactic PS (sPS). Metallocene-produced sPS is a semicrystalline material with a T_m of 520 °F. It was initially produced by Dow in 1997 under the tradename of Questra. It has good chemical and solvent resistance in contrast with "regular" PS, which has generally poor chemical and solvent resistance because of the presence of voids, due to the presence of the bulky phenyl groups, that are exploited by the solvents and chemicals.

Physical properties of PS are dependent on the molecular weight and presence of additives (Table 4.8). While higher-molecular-weight PS offers better strength and toughness, it also offers poorer processability. Low-molecular-weight PS allows for good processability but poorer strength and toughness. Generally a balance is sought where intermediate chain lengths are used.

Table 4.8 Properties of typical styrene polymers

Property	PS	HIPS	SAN	ABS
Glass transition temp. (T_g, °F)	212	203	248	239
Processing temp. (°F)	350	400	350	350
Molding pressure (10^3 psi)	15	15	15	20
Maximum resistance to continuous heat (°F)	175	175	200	180
Tensile strength (10^3 psi)	12	30	10	5
Flexural strength (10^3 psi)	15	50	12	10
Elongation (%)	2	40	2	20
Tensile modulus (10^3 psi)	400	250	550	350
Flexural modulus (10^3 psi)	425	200	500	350
Specific gravity	1.04	1.04	1.07	1.2

Because PS is brittle with little impact resistance under normal operating conditions, early work was done to impart impact resistance. The best-known material from this work is called high-impact polystyrene (HIPS). HIPS is produced by dispersing small particles of butadiene rubber in with the styrene monomer. Polymerization captures the butadiene rubber particles within the polymerizing styrene.

Major uses of PS are in packaging and containers, toys and recreational equipment, insulation, disposable food containers, electrical items and electronics, housewares, and appliance parts. Expandable PS is used to package electronic equipment such as TVs, computers, and stereos.

Legislation was put in place in some states to ensure the recycling of PS. Interestingly some of this legislation was written such that all PS had to be recycled with in some period of time such as a year. This legislation was changed to reflect the real concern of fast food containers (Figure. 4.10 and 4.11) when it was pointed out that less than 10% PS is used in this manner and that well over twice as much was used as house insulation that should not be recycled every year or so.

4.3.4 Poly(vinyl chloride)

Poly(vinyl chloride) (PVC) is one of the earliest produced polymers. In 1835 Justus von Liebig and his research student Victor Regnault reacted ethylene dichloride with alcoholic potash, forming the monomer vinyl chloride. Regnault believed he polymerized vinyl chloride, but later studies showed it to be poly(vinylidene chloride). In 1872, E. Baumann exposed vinyl chloride sealed in a tube to sunlight and produced a solid, PVC. Klasse, in Germany, found that vinyl chloride could be made by addition of hydrogen chloride to acetylene in a system that could be scaled up for commercial production. Today most of the vinyl chloride is made from the oxychlorination reaction with ethylene. By World War I, Germany was producing a number of flexible and rigid PVC products. One of its applications

Figure 4.10 Various PS foam sheet food packaging applications

Figure 4.11 Various oriented PS sheet food packaging applications

was a replacement for corrosion-resistant metals.

Waldo Semon was responsible for bringing many of PVC products to the market. The difficulty in fabricating PVC was associated with its tendency to decompose at temperatures typically used for molding and extrusion. Semon introduced materials called plasticizers, which allowed PVC to be processed at lower temperatures. As a young scientist at BF Goodrich, he worked on ways to synthesize rubber and to bind the rubber to metal. In his spare time he discovered that PVC, when mixed with certain liquids, gave an elastic-like, pliable material that was rainproof and fire-resistant and did not conduct electricity. Under the tradename Koroseal, the rubbery material came into the marketplace, beginning about 1926, as shower curtains, raincoats, and umbrellas. During World War II it became the material of choice to protect electrical wires for the Air Force and Navy.

As a side note, there is today a debate concerning the use of chlorine-containing materials and their effect on the atmosphere. This is a real concern and one that is being addressed by industry. PVC and other chloride-containing materials have in the past been simply disposed of through combustion that often created unwanted hydrogen chloride. This practice has largely been stopped, but care should be continued to see that such materials are disposed of properly. Furthermore, simply outlawing of all chloride-containing materials is not possible or practical. For instance we need common table salt for life, and common table salt is sodium chloride. Chlorine is widely used as a water disinfectant both commercially (for our drinking water) and for pools. Furthermore, PVC is an important material that is not easily replaced. Finally, the amount of chloride-containing residue that is introduced into the atmosphere naturally is large in comparison to that introduced by PVC. Even so, we must exercise care because we want to leave a better world for our children and grandchildren, so a knowledge-based approach must be taken.

Because of its versatility, some unique characteristics, ready availability and low cost, PVC is now the second largest produced synthetic polymer behind polyethylene. PVC materials are often defined to contain 50% or more by weight vinyl chloride units. PVC is generally a mixture of a number of additives and often other units such as ethylene, propylene, vinylidene chloride, and vinyl acetate. In comparison to many other polymers, PVC employs an especially wide variety of additives. For instance, a sample recipe or formulation for common stiff PVC pipe such as used in housing and irrigation applications may contain (in addition to the PVC resin) tin stabilizer, acrylic processing aid, acrylic lubricant-processing aid, acrylic impact modifier, calcium carbonate, titanium dioxide, calcium sterate, and paraffin wax. Such formulations vary according to the intended processing and end use. In such nonflexible PVC materials the weight amount of additive is on the order of 5% — 10% in mass.

PVC is used in a wider range of applications than any other plastics. From the home and office to the factory and leisure center, in virtually all areas of human activity (Figure 4.12). As one of the world's most versatile thermoplastics, with a wider range of uses than any other plastic material, such widespread use and acceptance is not surprising. PVC is versatile, durable, hygienic, safe and cost-effective. As such, it has significant advantages over other materials.

PVC is fundamentally a long lasting material with over 60% of its applications having a lifetime of more than 40 years. Around two thirds of all production is used in the building and construction sector. Other using significant volumes of PVC are the packaging industry for food, non-food and pharmaceutical products, in wire, cable and electrical applications and in the leisure and transport industries.

In more than 60 years of production and commercial use, PVC has contributed greatly to the development of modern life. It has also become one of the most thoroughly researched materials. Its manufacture is closely regulated and results of life cycle analyses published so far show that PVC is a safe product in its production, use and disposal.

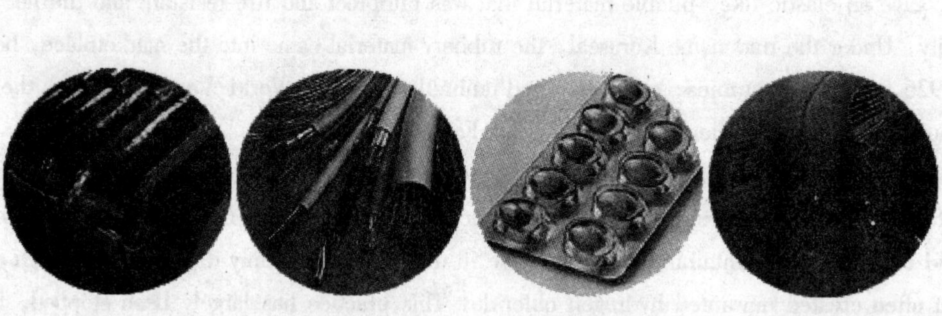

Figure 4.12 Applications of PVC in construction, electrical, medical and vehicles

4.4 Engineering plastics and thermosets

The most common engineering thermosets are materials such as the phenol and amino plastics that are crosslinked three-dimensional networks. These materials are not covered in this section. Here we will be looking at only engineering thermoplastics. As noted in section 4.3, the line between engineering thermoplastics and simply thermoplastics is not clear and varies with application and the person making the distinction.

The terms high-performance, engineering, and advanced thermoplastics are often used interchangeably. They generally contain no or only light crosslinking.

Engineering thermoplastics are bulk materials that can be easily and readily machined, milled, drilled, or otherwise have its shape modified while remaining in the solid state — much like metals. They retain their mechanical functionality even when subjected to mechanical stress/strain, vibration, friction, flexure, and so on over a general temperature range of 0 — 100 °C. They deform when the applied force is too great, yielding and deforming rather than simply cracking or breaking in two. This property is called impact resistance, that is, the ability to withstand shock without undergoing brittle failure. These materials are rapidly replacing metals because they offer advantages such as lightness per strength, corrosion resistance, self-lubrication, economy in fabrication, and in some cases transparency and ease in decoration. The need for engineering plastics is increasing. In 2001, more than 15 million tons of engineering thermoplastics were used in the United States.

4.4.1 Nylons

Nylons are one of the most common polymers used as fiber. Nylon is found in clothing all the time, but also in other places, in the form of a thermoplastic. Nylon's first real success came with its use in women's stockings (Figure 4.13), in about 1940. They were a big hit, but they became hard to get, because the next year the United States entered World War II, and nylon was needed to make war materials, like parachutes and ropes. But before stockings or parachutes, the very first nylon product was a toothbrush with nylon bristles.

Figure 4.13 Giant leg, thirty-five feet high, advertises nylons to the Los Angeles area (1939)

Nylons are also called polyamides, because of the characteristic amide groups in the backbone chain. Proteins, such as the silk nylon was made to replace, are also polyamides. These amide groups are very polar, and can form hydrogen bond with each other. Because of this, and because the nylon backbone is so regular and symmetrical, nylons are often crystalline, and make very good fibers. The nylon as shown in Figure 4.14 is called nylon 66, because each repeat unit of the polymer chain has two stretches of carbon atoms, each being six carbon atoms long. Other nylons can have different numbers of carbon atoms in these stretches.

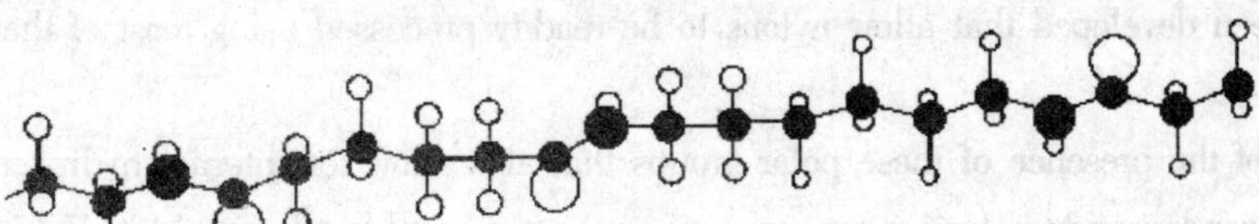

$$\left[\overset{\text{O}}{\overset{\|}{\text{C}}} - CH_2 - CH_2 - CH_2 - CH_2 - \overset{\text{O}}{\overset{\|}{\text{C}}} - NH - CH_2 - CH_2 - CH_2 - CH_2 - CH_2 - CH_2 - NH \right]_n$$

amide group

Figure 4.14 Ball-and-stick model and chemical structure of nylon 66

Nylons can be made from diacid chlorides and diamines. Nylon 66 is made from the monomers adipoyl chloride and hexamethylene diamine. This is one way of making nylon 66 in the laboratory. But in a nylon plant, it's usually made by reacting adipic acid with hexamethylene diamine:

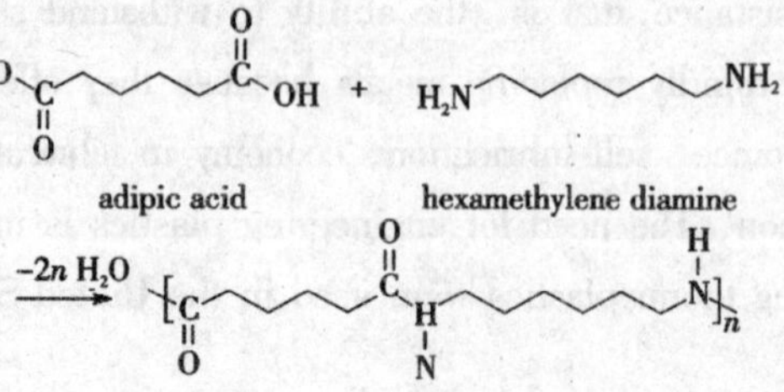

Another kind of nylon is nylon 6. It's a lot like nylon 66 except that it only has one kind of carbon chain, which is six atoms long.

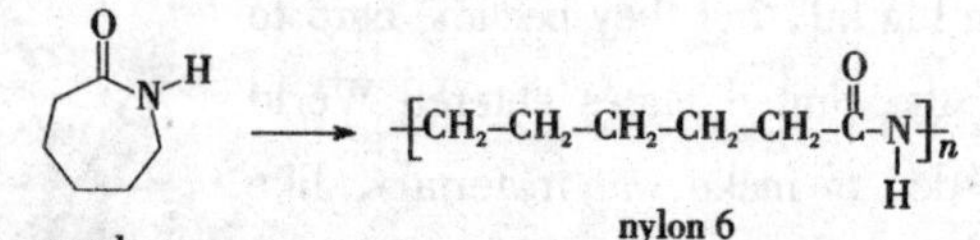

It's made by a ring opening polymerization from the monomer ε-caprolactam. Nylon 6 doesn't behave much differently from nylon 66. The only reason both are made is because DuPont patented nylon 66, so other companies had to invent nylon 6 inorder to get in to the nylon business.

Nylons offered new challenges to the chemical industry. Because of the presence of polar groups, the attractive forces between chains were high in comparison to vinyl polymers. Nylons are generally semicrystalline, meaning they have a good amount of order. Thus, while they have a T_g, the main physical transition is the T_m so that they undergo a sharper transition from solid to melt in comparison to many of the vinyls. Thus, the processing temperature window is more narrow. If melt flow is required for processing, then the temperature must be high enough to allow for ready flow but low enough so as not to break primary bonds with in the processed material. Even so, processing techniques have been developed that allow nylons to be readily processed using most of the standard techniques.

Because of the presence of these polar groups that also allow for internal hydrogen bonding, nylons and most condensation polymers are stronger, more rigid and brittle, and tougher in comparison to most vinyl polymers. Nylons are also "lubrication-free," meaning they do not need a lubricant for easy mobility so that they can be used as mechanical bearings and gears without need for periodic lubrication.

In general, more crystalline nylons are fibrous, whereas less crystalline nylon materials are more plastic in behavior. The amount of crystallinity is controlled through a variety of means including introduction of bulky groups and asymmetric units, rapid cooling of nonaligned melts, and introduction of plasticizing materials.

A number of aromatic polyamides, aramids, have been produced that are strong, can operate under high temperatures, and have good flame-retardant properties. Nomex™ is used in flame-resistant clothing and in the form of thin pads to protect sintered silica-fiber mats from stress and vibrations during the flight of the space shuttles. Kevlar™ is structurally similar, and by weight it has a higher strength and modulus than steel and is used in the manufacture of so-called bullet resistant clothing. Because of its outstanding strength/weight ratio, it was used as the skin covering of the Gossamer Albatross, which was flown over the English Channel using only human power.

Nomex Kevlar

4.4.2 Polyesters

Carothers and his research group at DuPont began to investigate the formation of polymers from the reaction of aliphatic diacids with diols, generally adipic acid and ethylene glycol (derived from reaction of ethylene oxide with water; major ingredient in most antifreeze), in search of materials that would give them fibers. They were only able to form syrupy mixtures. This is because unlike reactions with diamines, the equilibrium reaction greatly disfavors ester formation. Furthermore, the ability to have almost equal amounts of functional groups is easily achieved with the amines through formation of salts with the amines, but diols do not form such salts. Carothers' group understood the principle of "driving" an equilibrium reaction by removing water, thus forcing the reaction toward esterformation.

The initial polyester formation actually occurred much earlier and is attributed to Gay Lussac and Pelouze in 1833 and Berzelius in 1847. These polyesters are called glyptals and alkyds, and they are useful as coatings materials and not for fiber production. While these reactions had low fractional conversions, they formed high-molecular-weight materials because they had functionalities (that is, number of reactive groups on a single reactant) greater than two, resulting in crosslinking.

The heat resistance of Carothers' polyesters was not sufficient to withstand the temperature of the hot ironing process. Expanding on the work of Carothers and Hill on polyesters, Whinfield and Dickson, in England, overcame the problems of Carothers and co-workers by employing an ester interchange reaction between ethylene glycol and the methyl ester of terephthalic acid, forming the polyester, poly(ethylene terephthalate) (PET), with the first plant coming on line in 1953. This classic reaction producing Dacron, Kodel, Terylene fibers and Dacron fibers. Methyl alcohol, methanol, is lower boiling than water (149 °F compared with 212 °F) and thus more easily removed, allowing the reaction to be forced toward polymer formation more easily.

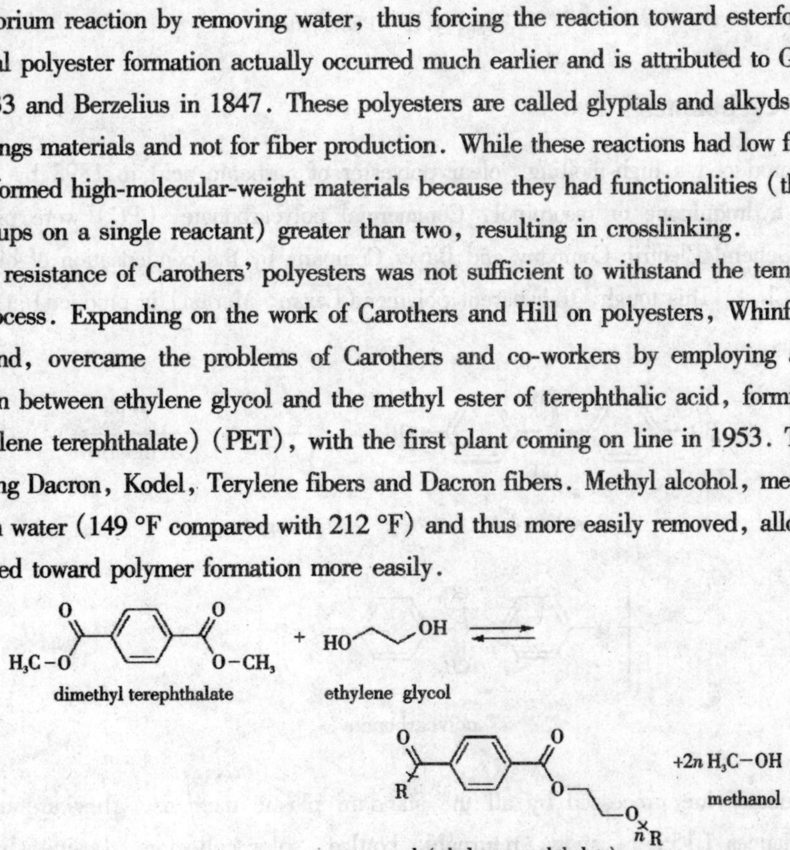

PET is now extensively used as bottling material for soft drinks because of its low carbon dioxide permeability (Figure 4.15). Carbon dioxide permeability decreases with increasing film thickness and crystallinity. To achieve the necessary crystallinity, partially crystalline PET is employed in the stretch blow molding process with the molding process carried out to promote further crystalline formation. It is also used for molded automobile parts. Over 500,000 tons of polyester engineering plastics are produced annually in the United States.

Figure 4.15 PET bottles

4.4.3 Polycarbonates

Einhorn produced a high-melting, clear polyester of carbonic acid in 1898 by the reaction of phosgene and hydroquinone or resorcinol. Commercial polycarbonates (PC) were produced in the 1950s by the General Electric Company and Bayer Company by the condensation of bisphenol A and phosgene ($COCl_2$). This tough, transparent polymer (Lexan, Merlon) is produced at an annual rate of 130,000 tons.

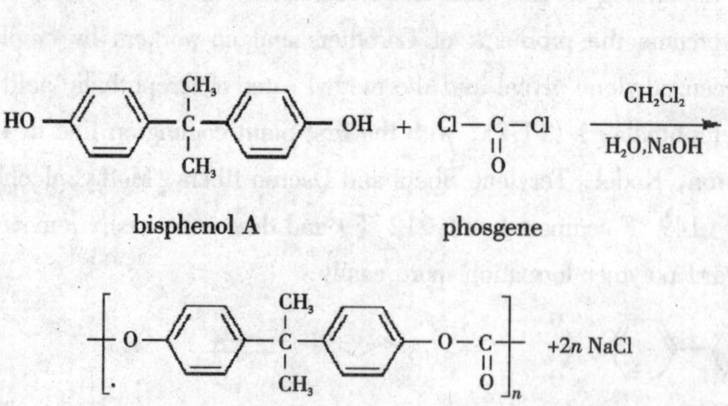

Polycarbonates are processed by all the standard plastic methods. They are used in glazing (40%), appliances (15%), signs, returnable bottles, solar collectors, business machines, and electronics. They show good creep resistance, good thermal stability and a wide range of use temperatures (about − 60 °F to 270 °F). Coatings are generally used on PC sheets to improve mar and chemical resistance.

Nonrecordable compact discs (CDs) are made of rigid, transparent polycarbonates with a reflective metal coating on top of the polycarbonate. A laser is used to encode information through creation of physical features sometimes referred to as "pits and lands" of different reflectivity at the polycarbonate-metal interface.

Recordable CDs contain an organic dye between the polycarbonate and metal film. Here, a laser creates areas of differing reflectiveness in the dye layer through photochemical reactions. A beam from a semiconductor diode laser "interrogates" the undersides of both types of CDs seeking out areas of reflected light, corresponding to the binary "one", and unreflected light, corresponding to the binary "zero". The ability to "read" information is dependent on the wavelength of the laser. Today, most of the CD players use a near-infrared laser because of the stability of such lasers. Efforts are underway to develop stable and inexpensive lasers of shorter wavelengths that will allow the holding of more information within the same space.

4.4.4 Polysiloxanes

Polysiloxanes, also called silicones, offer a combination of electrical, chemical and mechanical properties not found for any other class of polymers. They show relatively high oxidative and thermal stability, low power loss, and unique flow and stress/strain properties, are inert to most ionic and inert reagents, exhibit low flow changes as temperature varies, are nonflammable, and have good shear stability, high compressability, low surface tension, and so on. They have an exceptionally wide use temperature from about − 120 °C to 200 °C or an operating temperature range of about 300 degrees. This temperature range allows their use in extreme temperatures. The first footprints on the moon were made with polysiloxane elastomeric boots.

The first polysiloxanes were unstable, but this instability was overcome by capping the end groups. Almost all of the commercial polysiloxanes are based on polydimethylsiloxane, with trimethylsiloxy end groups.

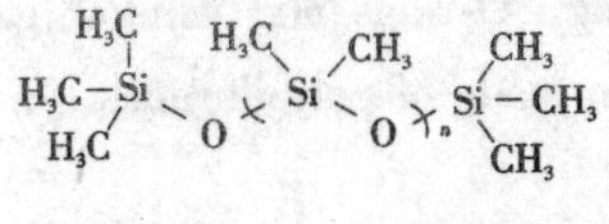

The reason for the low-temperature flexibility is because of a very low T_g, about − 120 °C, which is the result of the methyl groups attached to the silicon atoms being free to rotate, causing the oxygen and other surrounding atoms to "stay away" creating a flexible chain.

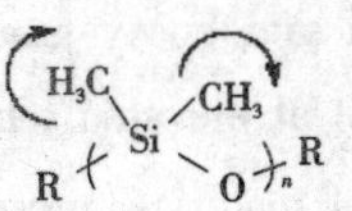

The viscosity or resistance to flow increases as the number of repeat units increase, but physical properties such as surface tension and density remain about the same after a degree of polymerization

(DP) of about 25. The liquid surface tension is lower than the critical surface tension of wetting, resulting in the polymer spreading over its own absorbed films. The forces of attraction between polysiloxane films are low, resulting in the formation of porous films that allow oxygen and nitrogen to readily pass through, but not water. Thus, semipermeable membranes and films, have been developed that allow drivers to "breath air under water" for short periods.

As noted above, viscosity increases with DP, allowing many of the uses to be grouped according to chain length. Low-viscosity fluids with DP values of 2 — 30 are used in antifoams and in the flow control of coating applications. These applications are the direct consequence of the low attractions between polysiloxane chains, which, in turn, are responsible for their low surface tension. Thus, they encourage a coatings material to flow across the surface, thereby filling voids, corners, and crevices. Their good thermal conductivity and fluidicity at low temperatures allows their use as low-temperature heat exchangers and in low-temperature baths and thermostats.

Viscous fluids correspond to a DP range of about 50 — 400. These materials are employed as mold release agents for glass, plastic and rubber parts. They are good lubricants for most metal to non-metal contacts. They are used as dielectric fluids (liquids) in a variety of electrical applications including transformers and capacitors; as hydraulic fluids in vacuum and hydraulic pumps; in delicate timing and photographic devices; as antifoam agents; as components in protective hand creams; as toners in photocopiers; in oil formulations when mixed with thickeners; and in inertial guidance systems. High-performance greases are formed by mixing the polysiloxane fluids with polytetrafluoroethylene or molybdenum disulfide. Brake fluids are formulated from polydimethylsiloxane fluids with DP values of about 50. High-viscosity fluids with DP values of about 700 — 6,000 are used as damping fluids for weighting meters at truck stops. They act as liquid springs in shock absorbers. The longer-chained fluids are used as impact modifiers for thermoplastic resins and as stationary phases in gas chromatography.

As with the alkanes, even longer chains form the basis for solid polysiloxanes that according to design can be classified as thermoplastics, engineering thermoplastics, elastomers and (when crosslinked) thermosets. Solid polysiloxanes are used in a variety of applications including sealants, thermostripping, caulking, dampening, O-rings and window gaskets. Weather stripping on cooling units, trucks and automobiles is often made of polysiloxanes.

4.4.5 Epoxy resins

The word epoxy is derived from the Greek prefix epi, meaning between, and the English suffix of oxygen. Epoxy resins, which are obtained by the condensation of bisphenol A and epichlorhydrin, were described by Linderman in the last part of the nineteenth century and patented by P. Schlack in 1934. These linear prepolymers contain hydroxyl and epoxy groups, which may be crosslinked by reaction with anhydrides, such as maleic anhydride at elevated temperatures, or with amines [$R(NH_2)_n$] at moderate temperatures. By varying ratios of materials present in the prepolymers, large differences in curing rate and hardness are possible.

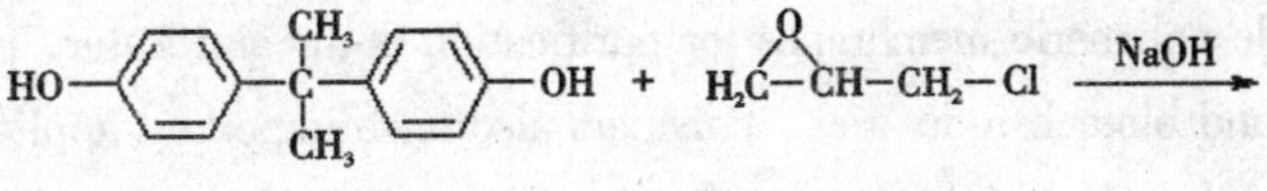

bisphenol A epichlorhydrin

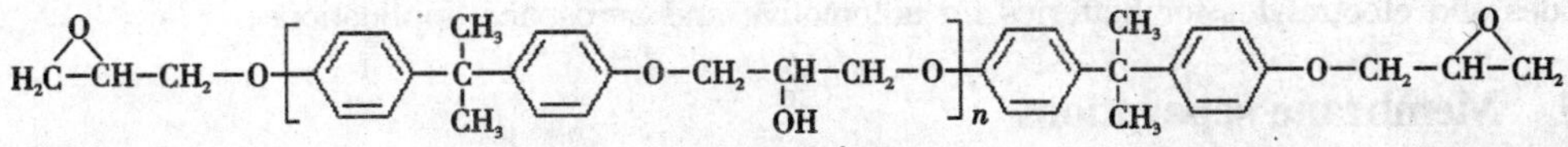

epoxy resin

Over 225,000 tons of epoxy resins are used annually in the United States as adhesives, coatings and encapsulating compositions. Fiberglass-reinforced epoxy resins are used as aircraft components. About 110,000 tons of epoxy resins are used annually as coatings, and 45,000 tons for fiber-reinforced plastics. The wide range of applications of epoxy resins is a result of the versatility of the system.

There are two major types of surface coatings. The first is ambient-temperature-cured. These are crosslinked using polyamides and polyamines. The second type of coating is heat-cured. These are cured using formaldehyde resin, anhydrides and polycarboxylic acids. Phenol-formaldehyde resin-epoxy products include drum and tank linings, wire coatings, impregnation varnishes, and food and beverage can coatings. Epoxy coatings can be used as powder coatings, thus eliminating the use of solvents. Printed circuit boards are made from fiberglass laminates and epoxy resins. Epoxy resins are used as binders for floor surfaces that are subject to heavy traffic, in patching concrete, and in casting, encapsulation, and potting electrical equipment. The common adhesive sold in hardware stores is generally a two-component liquid or paste that cures to give an epoxy resin. Commercial epoxy resins and their structural applications are shown in Figure 4.16.

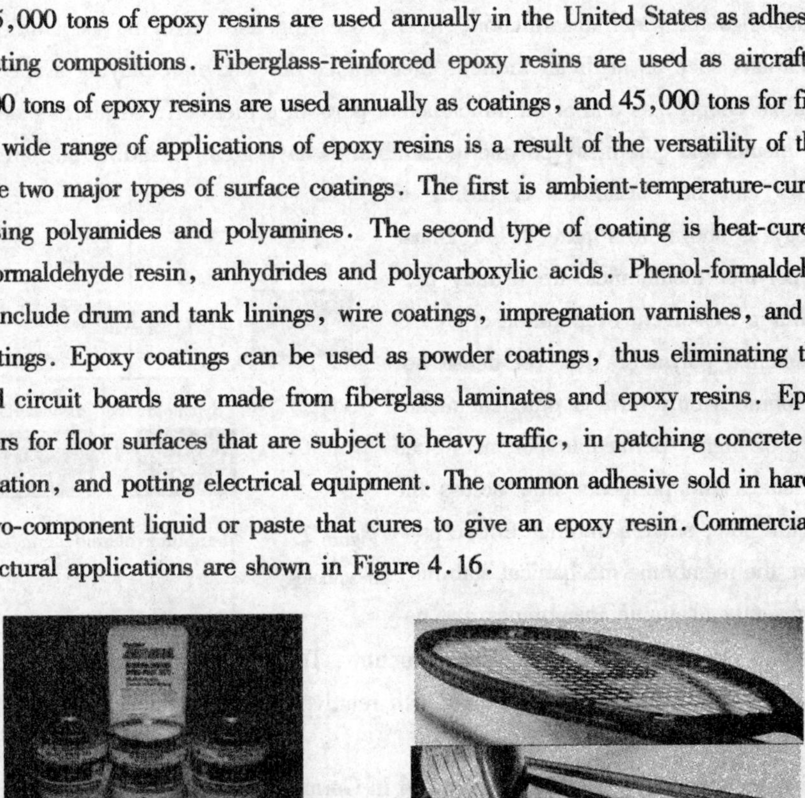

(a) (b)

Figure 4.16 Epoxy resins (a) and their structural applications (b)

4.5 Applications of polymers

As discussed earlier sections, plastics are important materials for use in commodity products such as textiles, tires, and packaging (e.g., film and containers). Polymers, particularly thermosets, also have widespread use as composite materials for applications in transportation, including automotive, marine, and aerospace materials. Most of the markets are now mature and future growth is expected to be small; however, there are many more applications for polymers that may be less obvious but have great potential and offer challenges for new technology and growth in the twenty-first century. Some of these uses include polymeric membranes for purification of air and water, and for important separations in the chemical and biotech industries. Polymers also have important applications in medicine, including uses in controlled drug delivery, artificial organs, and protein synthesis. Some polymers can be made to be electrically conductive and offer potential for the semiconductor industry and as lightweight electrodes and electrolytes for batteries for automotive and aerospace applications.

4.5.1 Membrane separations

Membrane technology has gained a huge importance in the last 30 years, competing with long established technologies for water desalination, food processing and emerging as a unique solution in medical applications such as artificial kidney. Membranes are also now playing a special role in the field of alternative energy, as one of the fundamental parts of a fuel cell. Although membranes can be prepared from metals and ceramics, polymeric membranes have great versatility and are widely used.

Membranes can be classified, according to their morphology as shown in Figure 4.17. Dense homogeneous polymer membranes are usually prepared from solution by solvent evaporation or by extrusion of the melted polymer. However dense homogeneous membranes only have a practical meaning when made of highly permeable polymers such as silicone. Usually the permeant flow across the membrane is quite low, since a minimal thickness is required to give the membrane mechanical stability. Most of the presently available membranes are porous or consist of a dense top layer on a porous structure. The preparation of membrane structures with controlled pore size involves several techniques with relatively simple principles, but which are quite tricky.

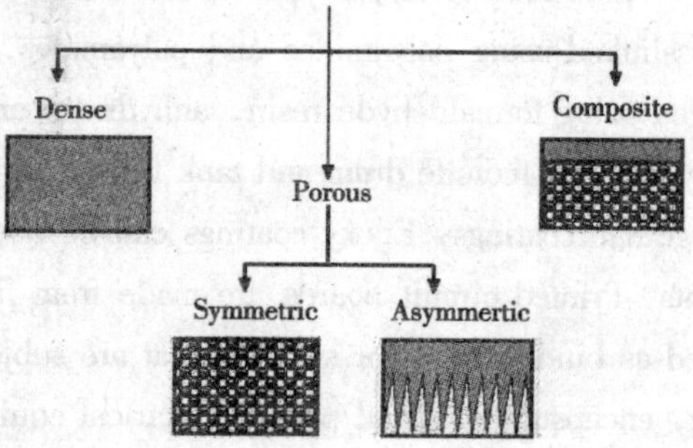

Figure 4.17 Membrane classification according to the morphology

Commercial membranes were already produced in Germany by Sartorius in the early 1920s. However they had only a limited application on a laboratory scale. The break-through of the membrane technology came first in the 1960s with the development of the asymmetric porous membranes by Loeb and Sourirajan. The asymmetric membranes combine high permeant flow, provided by a very thin selective top layer and a reasonable mechanical stability, resulting from the underlying porous structure. An asymmetric structure characterizes most of the presently commercially available membranes, which are now produced from a wide variety of polymers. By far the most common method used in generation of asymmetric structures in membranes is the "phase inversion" process. Particularly in the case of microfiltration, several techniques other than phase inversion are currently applied in the industry.

The phase inversion process consists of the induction of phase separation in a previously homogeneous polymer solution either by temperature change, by immersing the solution in a non-solvent bath (wet process) or exposing it to a non-solvent atmosphere (dry process).

In the thermal process, a low molecular weight component usually acts as a solvent at high temperature and as a non-solvent at low temperature. It is then removed after formation of the porous structure. Although the thermal process can be applied to a wide range of polymers, it is especially interesting for those with poor solubility, such as polypropylene, which can be hardly manufactured into a porous membrane by other phase inversion processes. An isotropic microporous structure is usually formed.

The isothermal phase inversion is commercially more widespread. Usually the polymer solution is immersed in a non-solvent bath (wet process) and a solvent-non-solvent exchange leads to phase separation. The polymer rich phase forms the porous matrix, while the polymer poor phase gives rise to the pores. The morphology is usually asymmetric, with a selective skin on the surface, as shown in Figure 4.18.

Figure 4.18 Asymmetric porous membrane

Reverse osmosis

At the extreme limit of filtration is a process called reverse osmosis, which is used to obtain potable water from brackish water or salt water. The process is termed "reverse" osmosis since water would be expected to diffuse from the side of low salt concentration to that of high concentration in normal osmosis.

The most common membrane materials for reverse osmosis membranes are cellulose acetate, polyamide and the "thin film composites" prepared by interfacial polymerization on the surface of a porous support. Cellulose acetate (CA) was one of the first membrane materials, and it is still being successfully used, especially in water treatment. They usually allow quite high water flows with low salt solubility. One advantage, when compared to the polyamide based membranes, is its chlorine tolerance. Also because of the neutral surface, cellulose acetate membranes usually exhibit a more stable performance than polyamide membranes in applications where the feed waters has a high fouling potential, such as in municipal effluents and surface water supplies. However, CA membranes are drastically less stable in organic solvents than polyamide. The recommended pH range is between 3 and 7, they are less resistant to biological attack and the recommended temperature is lower than 50 °C. The susceptibility to hydrolysis increases with temperature and is an inverse function of the degree of acetylation. Aromatic polyamides have a much higher solvent resistance and may be used in a wider pH range (pH 4 — 11). The main application is the treatment of brackish water and seawater. They can be produced in very thin hollow fibers with large surface area/volume. The membrane top layer is however quite thick (>0.2 μm), which leads to relatively low water flows. The main disadvantage is the very low chlorine tolerance.

Nanofiltration

Another growing area in membrane filtration is nanofiltration. The performance of nanofiltration membranes falls between reverse osmosis and ultrafiltration membrane with pore sizes in the area of 1 nm and nominal molecular-weight cutoffs in the range from 100 to 200. Nanofilters are usually prepared as thin-film composite membranes consisting of a thin separating layer containing negatively charged hydrophobic groups attached to an ultrafiltration membrane support. Polymers suitable for the separating layer include cellulose acetate, polyamides, poly(vinyl alcohol), sulfonated polysulfone and polyethersulfone. As illustrated by Figure 4.19, nanofiltration membranes retain sugars and some multivalent salts (e.g., $MgSO_4$) but pass many monovalent salts (e.g., NaCl) and undissociated acids. Salt rejection is due to electrostatic interactions between ions and the separating layer. Applications for nanofiltration membranes include the demineralization of water, the removal of heavy metals, and the removal of lignin and related impurities form weed-pulp streams. Advantages of nanofiltration membranes over reverse osmosis membranes are higher water flux and improved fouling resistance against hydrophobic colloids, proteins, and oils.

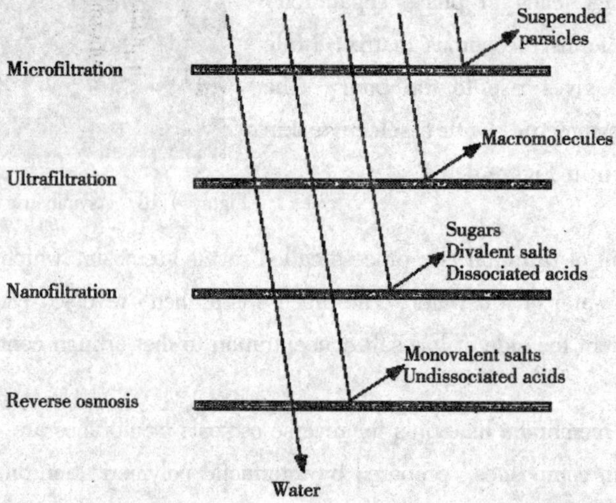

Figure 4.19 Comparison of the rejection performance of microfiltration, ultrafiltration, nanofiltration and reverse osmosis

Ultrafiltration

The most widely used polymer for the preparation of ultrafiltration membranes is polysulfone (PSF) or polyethersulfone (PES).

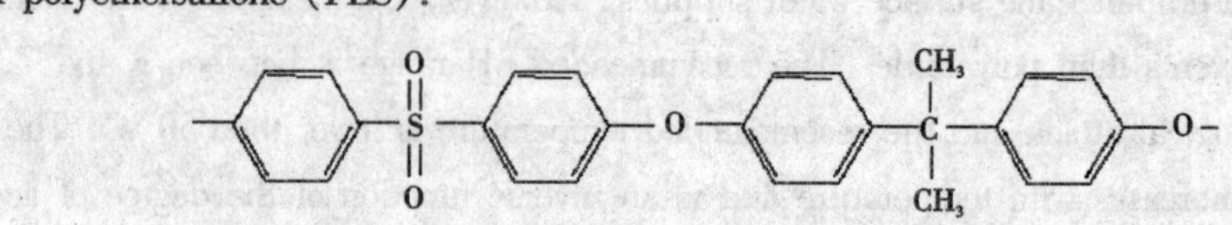

polysulfone

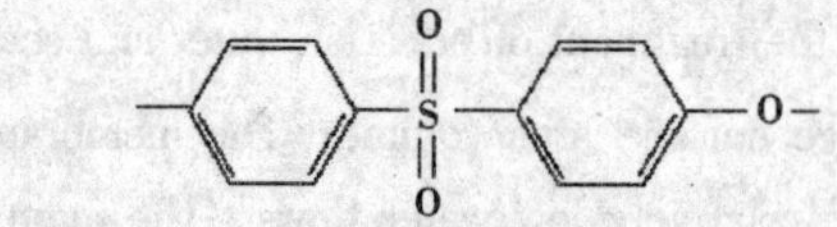

polyethersulfone

The first developments of PSF membranes appeared in 1960s as an alternative to cellulosic membranes. Since then several procedures have been described in the literature for PSF membranes, in many cases using the high molecular weight polysulfone Udel P-3500 commercialized by Amoco. A great advantage when compared to cellulose acetate is its resistance in extreme pH conditions, as well as its thermal stability. PSF has a T_g of 195 °C and PES even higher, at 230 °C. Both PSF and PES are soluble in chloroform and *N*,*N*-dimethyl formamide (DMF), and are easily applied in conventional phase inversion processes. This high solubility is also the main drawback of PSF as a membrane material, eliminating the use of polysulfone-supported membranes in the processing of solvent-based feed solutions. It is also a problem coating the PSF support with polymers, which are only soluble in organic solvents. Another disadvantage of PSF and PES membranes is their hydrophobic character, which prevents spontaneous wetting with aqueous media. Consequently, the membrane must be prevented from drying completely or the membrane must be treated with a hydrophobic agent, glycerin, for example, before drying. Another serious disadvantage of hydrophobic materials consists in the fact that they often possess a powerful nonspecific adsorption capacity. This phenomenon, known as fouling, leads to a rapid deterioration of the membrane permeability. Suggestions for hydrophilic membranes have already been proposed that do not suffer from these disadvantages. Several procedures have been proposed to make membrane surface more hydrophilic.

Poly(vinylidene fluoride) (PVDF) is quite interesting in the manufacture of ultrafiltration membranes due to its chemical resistance. PVDF is resistant to most inorganic and organic acids and can be used in a wide pH range. It is also stable in aromatic hydrocarbons, alcohols, tetrahydrofurane and halogenated solvents. Furthermore it is resistant to oxidizing environments including ozone, which is used in water sterilization. PVDF is semi-crystalline with a very low T_g (− 40 °C), which makes it quite flexible and suitable for membrane application in temperatures ranging between − 50 °C and 140 °C, just prior to its melting temperature. Although stable in most of organic solvents, PVDF is soluble in DMF, *N*, *N*-dimethyl acetamide (DMAC), *N*-methyl pyrrolidone (NMP) and dimethylsulfoxide, making membrane preparation by phase inversion possible. The morphology of a PVDF membrane obtained from a solution in DMAC is shown in Figure 4.20.

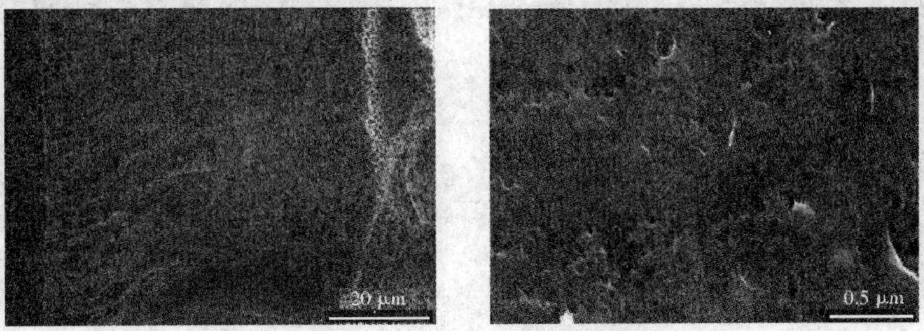

Figure 4.20 Poly(vinylidene fluoride) membrane: (left) cross-section and (right) surface

Polyacrylonitrile (PAN) has been used in the preparation of ultrafiltration membranes for a long time due to its superior resistance to hydrolysis and oxidation. PAN is highly crystalline and relatively hydrophilic and is usually copolymerized with more hydrophilic monomers to improve processability and to make it less brittle. Hollow fibers can be prepared from PAN dissolved in nitric acid. Preparation of PAN membranes by phase inversion from solutions in DMAC, DMF or NMP is also possible. In order to make the membranes suitable for reverse osmosis, they were submitted to a plasma treatment in the presence of 4-vinyl pyridine.

Microfiltration

Several of the polymers mentioned above for ultrafiltration can also be used for the preparation of microfiltration membranes by phase inversion. However, for the range of pore size useful for microfiltration, other procedures have also been successfully used for the preparation of commercial membranes. Among these are stretching, track etching, leaching and sintering.

Stretching is part of the preparation process both of the Celgard® and the Gore-Tex® membranes. Cold drawing has been described already in 1969 for membrane preparation starting from crystalline polymers. Another preparation method is solvent stretching, where the precursor film is brought in contact with a swelling agent and stretched. The swelling agent is removed while the film is maintained stretched to render the film microporous. Other processes use sequencial "cold" and "hot" stretching steps.

The Celgard® membrane is made of polypropylene, which is a low cost and quite inert polymer. It is resistant under extreme pH conditions and is insoluble in most solvents at room temperature. It swells however in very apolar solvents such as carbon tetrachloride. No solvent is required for the preparation of the membrane. It involves the extrusion of PP films with high melt stress to align the polymer chains and induce the formation of lamellar microcrystallites when cooling. The film is then 50% — 300% stretched just below the melting temperature. Under stress, the amorphous phase between the crystallites deforms, giving rise to the slit-like pores of the Celgard® membrane (Figure 4.21). The film is then cooled under tension. PP is highly hydrophobic and several surface treatments have been proposed to improve the hydrophilicity of PP membranes. Incorporation of surfactants is used to make Celgard® membranes more hydrophilic.

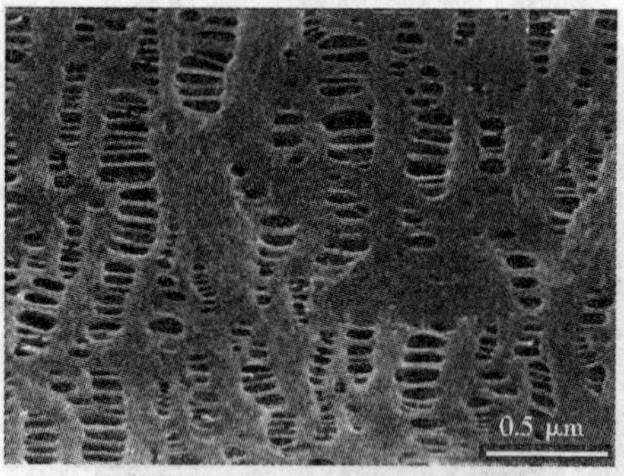

Figure 4.21 Celgard® membrane

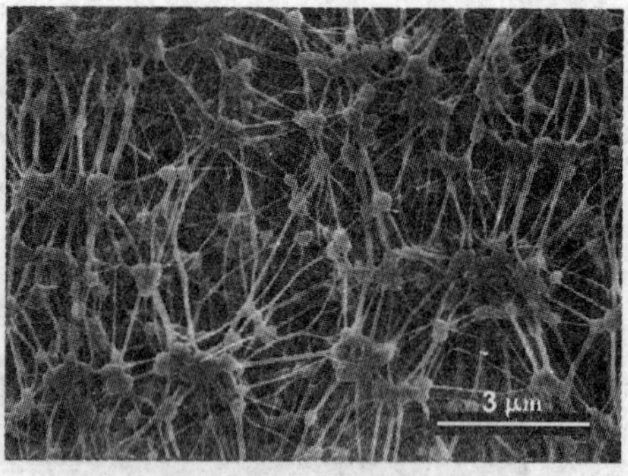

Figure 4.22 Gore-Tex® membrane

Another commercial membrane prepared by stretching is Gore-Tex (Figure 4.22). The polymer here is poly(tetrafluoroethylene) (PTFE), what makes the membrane extremely inert and thus convenient for processing even harsh streams. Processing PTFE is only possible by paste extrusion. In paste forming the polymer is mixed with a lubricant such as odorless mineral spirits naphtha or kerosene. The lubricant component is removed by heating up to 327 °C. Above this temperature, sintering would lead to a dense PTFE film. After lubricant removal, the PTFE film is submitted to a uniaxial or biaxial stretching, giving rise to an interconnected pore structure. The process was proposed by Gore and the resulting porous film is today a successful product in the membrane and textile industry. For uniaxial stretching the unsintered film from the paste extrusion is fed to a machine with heating roles, where one role is driven faster than the previous one to input stress and induce the pore formation. The difference in speed determines the amount of stretch. Additionally in the Gore patent a biaxial stretching is performed using a pantograph.

A special characteristic of the Gore membrane is that, since PTFE is very hydrophobic, (liquid) water must not be allowed to wet the membrane and its transport is hindered. On the other hand, water vapor can freely pass through the micropores, making the film suitable for transpirating impermeable cloths. However due to their inertness, PTFE membranes are also interesting during the processing of aggressive streams. If the goal is the filtration of aqueous wastes the membranes should be modified to become hydrophilic. A solution is disclosed in a later Gore patent, mixing PTFE with silica and a surfactant to form the paste for the extrusion. To improve the adhesion of the filler to the PTFE matrix, the membrane, in the last step of preparation, is heated with a solution of dimethyl octadecyl chlorosilane in toluene.

4.5.2 Biomedical applications

Polymers are widely used in medicine as biomaterials for sutures, artificial organs such as kidney and heart, orthopedic implants, tissue engineering scaffolds, and devices for the controlled release of drugs. Depending upon the application, polymers used in the biomedical field, must have one or more specific properties such as biocompatibility, selective permeability, the ability to biodegrade, and high strength or modulus. A few examples of the biomedical uses of polymers are given in this section.

Kidney dialysis

One of the important medical applications of polymeric membranes is kidney dialysis, which is used to remove toxins of low to moderate molecular weight from blood. These include sodium chloride, potassium chloride, urea, creatinine, and uric acid. At the same time, larger molecules necessary for life, such as cellular particles and plasma proteins, must be retained. The driving force for dialysis is a concentration gradient. Polymers suitable for kidney dialysis include cellophane (regenerated cellulose), which can be prepared in the form of small-diameter (ca. 200 μm) hollow fibers approximately 17 cm in length. As many as 10,000 of these fibers may be used in a typical dialysis module. Practical problems that are encountered in operation include concentration polarization and clogged pores due to plasma proteins that limit performance.

Soft tissue replacement

The success of soft tissue implants has primarily been due to the development of synthetic polymers. This is mainly because the polymers can be tailor-made to match the properties of soft tissues. In addition, polymers can be made into various physical forms, such as liquid for filling spaces, fibers for suture materials, films for catheter balloons, knitted fabrics for blood vessel prostheses, and solid forms for cosmetic and weight-bearing application.

It should be recognized that different applications require different materials with specific properties, the following are minimal requirements for all soft tissue implant materials: reasonably close approximation of physical properties, especially flexibility and texture, no changes in properties after implantation with time, no adverse tissue reactions, sterilizable, noncarcinogenic, nontoxix, nonallergenic, nonimmunogenic and etc.

The most common soft tissue implants are sutures. In terms of their performance, suture materials are generally classified into two broad categories; absorbable and nonabsorbable. Absorbable suture materials lose their entire tensile strength within 2 to 3 months; those that retain their strength longer than 2 to 3 months are nonabsorbable. The absorbable suture materials (some shown in Figure 4.23) are catgut (collagen sutures derived form sheep intestinal submucosa), reconstituted collagen, polyglycolide (PGA), poly(glycolide-lactide) random copolymer (Vicryl®), poly-*p*-dioxanone (PDS®, PDSII®), poly(glycolide-trimethylene carbonate) block copolymer (Maxon®), poly(glycolide-ε-caprolactone) (Monocryl®), and glycolide-trimethylene carbonate block copolymer (Biosyn®), The nonabsorbable sutures are divided into natural fibers (i.e., silk, cotton, and linen), and man-made fibers (i.e., polyethylene, polypropylene, polyamide, polyester, polytetrafluoroethylene and stainless steel).

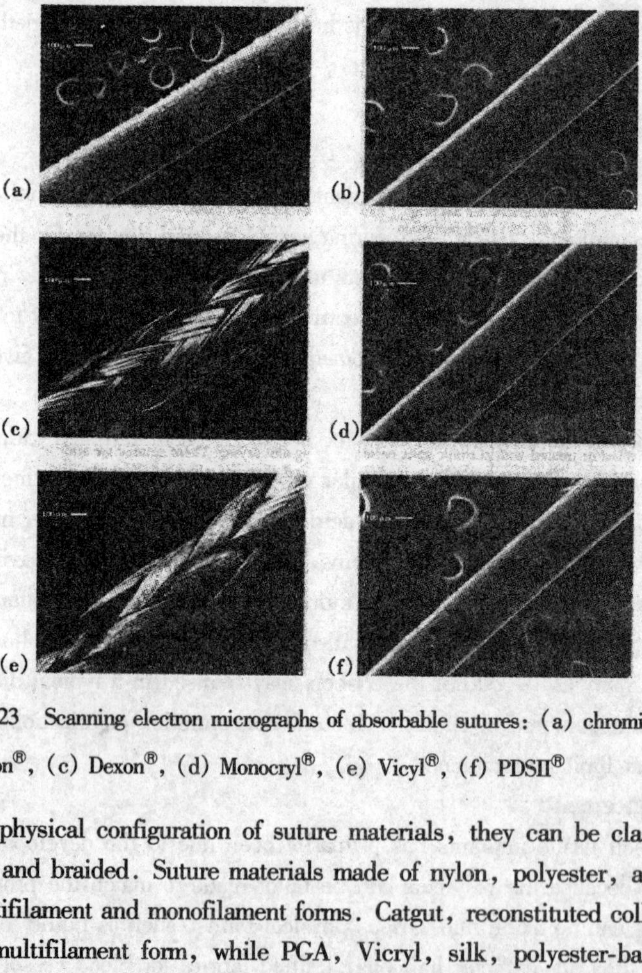

Figure 4.23 Scanning electron micrographs of absorbable sutures: (a) chromic catgut, (b) Maxon®, (c) Dexon®, (d) Monocryl®, (e) Vicyl®, (f) PDSII®

In terms of the physical configuration of suture materials, they can be classified into monofilament, multifilament, and braided. Suture materials made of nylon, polyester, and stainless steel are available in both multifilament and monofilament forms. Catgut, reconstituted collagen, and cotton are available on twisted multifilament form, while PGA, Vicryl, silk, polyester-based, and polyamide-based suture materials are available in the braided multifilament configuration. PDS, Maxon, Monocryl, Biosyn, polypropylene, and polytetrafluoroethylene suture materials exist in monofilament form only. Stainless steel metallic suture materials can be obtained in either monofilament or twisted multifilament configurations. Another unique physical configuration of suture material is available in polyamide (nylon 6) and has the trade name Supramid®, which has a twisted core covered by a jacket of the same material.

Suture materials are frequently coated to facilitate their handling properties, particularly a reduction in tissue drag when passing through the needle tract and the ease of sliding knots down the suture during knotting (i.e., knot tie-down). Although nonabsorbable beeswax, paraffin wax, silicone, and polytetrafluoroethylene are the traditional coating materials, new coating materials have been reported, particularity those that are absorbable sutures must be absorbable, and traditional nonabsorbable coating materials like wax are not appropriate for absorbable sutures. Furthermore, absorbable coating materials should have better tissue biocompatibility because of the lack of chronic tissue reaction.

The trend is to develop coating materials that have a chemical property similar to the suture to be coated. There are basically two types of absorbable coating materials: water soluble and insoluble. Water-insoluble coating materials have chemical constituents similar to those of the suture, and they are broken down by hydrolysis. They remain on the suture surface longer than water-soluble coatings. A typical example is polyglactin 370 used for Vicryl suture. Dexon II sutures have a polycaprolate coating that is water insoluble. Water-soluble coating materials dissolve promptly to reveal the uncoated suture after wound closure. A typical example is poloxamer-188 found on Dexon Plus. Multifilament sutures are more commonly provide with coating materials than monofilament sutures. For example, multifilament Vicryl and Dexon Plus or II have coating materials applied, while monofilament PDS and Maxon sutures have no coatings.

Controlled drug delivery

Medical drugs have a specific therapeutic range of concentration within which these are effective. Below this range, they are ineffective in the treatment of infection and disease and they can be toxic to healthy cells at higher concentrations. Obviously, it is desirable to maintain a constant level of drug concentration that is within the therapeutic range for an extended period of time. It is also important for effective drug therapy that the drug be able to target the specific area of disease in the body, especially in the administration of antitumor agents in the treatment of cancer. In these respects, conventional methods of drug administration, including pills and injections, are not satisfactory, and novel techniques for controlled drug delivery are being sought.

Controlled drug delivery provides two important potential applications for the use of polymers in the effective management of medical drugs in the body. The first of these is controlled release whereby a steady therapeutic concentration of the drug is maintained. Current examples of the use of controlled drug release include the delivery of contraceptives and the treatment of glaucoma. The other application is site-directed drug delivery whereby a polymer serves as a carrier to bring a drug to a specific site in the body, such as a site of infection, diseased organ, or collection of malignant cells.

The rate of drug release can be controlled by diffusion, reaction, or solvent. In diffusion control, the driving force for diffusion is the concentration gradient across the delivery device, which can be a reservoir or matrix system. In the former case, the reservoir containing the drug is encapsulated by a polymeric membrane, which can be either microporous or nonporous. Delivery is constant in time if the drug core is maintained at a saturated level. In the case of a matrix system, the drug is dissolved or dispersed in a polymer. In contrast to the reservoir system, the release rate in a matrix device will decrease with time as the distance the drug has to travel from within the matrix increases due to depletion of drug concentration at the surface of the device. Polymers that have been used for diffusion-controlled devices include polydimethylsiloxane and poly(ethylene-co-vinyl acetate).

Reaction-controlled systems utilize biodegradable polymers as a means of delivery. The drug can be physically dispersed in the polymer, which gradually degrades in the body as a result of hydrolysis or enzymatic attack. Alternately, the drug can be chemically linked to a polymer chain by a suitable spacer group that provides a biodegradable link. For reaction-controlled delivery, poly(lactide-co-glycolide) is particularly attractive because its breakdown products, lactic and glycolic acids, are biologically safe.

In solvent-controlled delivery, drug release is regulated by the permeation of water through the polymer. This may be achieved by the use of a simple osmotic pump. The drug is enclosed by a semi-permeable membrane that allows water to penetrate into the device due to osmotic pressure. The drug solution is dispersed through the orifice of the device at a rate equal to the rate of water uptake.

Another type of solvent-controlled system for controlled release uses polymeric hydrogels, which are water-swollen crosslinked polymer networks. The drug is incorporated in hydrogel polymer in the dry (glassy) state. When introduced in an aqueous environment, the polymer swells. The resulting gel provides a medium favorable for drug diffusion. A commonly used hydrogel is prepared from poly(2-hydroxyethyl methacrylate) (PHEMA) crosslinked by copolymerization with a bifunctional polymer. Since PHEMA is water soluble, the crosslinked network will swell to form a gel containing about 35% water in an aqueous environment. In addition to controlled-release applications, PHEMA is used in the manufactured of soft contact lenses.

Although the technology for site-directed drug delivery is still in its infancy, some examples illustrate its potential. One application uses a block copolymer of poly(ethylene glycol) (PEG) and poly (aspartic acid) as the drug carrier. PEG provides the carrier solubility, while the poly(amino acid) is used to attach an anticancer drug (e.g., adriamycin).

4.5.3 Applications in electronics

In general, polymers have very poor electrical conductivity. In fact, some polymers such as polytetrafluoroethylene find applications as electrical insulators. In the 1800s, it was observed that the conductivity of natural rubber — normally an excellent insulator — could be significantly increased by adding carbon black which has natural conductivity. In the 1930s, natural rubber filled with acetylene black was used as an antistatic device in hospitals and other facilities where it was necessary to control the danger of sparks due to the buildup of static electricity.

Before 1973, only one polymer — polysulfurnitride $(SN)_x$ — was known to have any appreciable conductivity (~ 10^3 S/cm). As in the case of natural rubber, it was found that this conductivity could be increased by doping with an electron acceptor such as bromine (~ 10^4 S/cm). Unfortunately, interest in this polymer was less than it would otherwise be because of the explosive nature of its precursor monomer, S_2N_2.

By the late 1970s, researchers in Japan and the United States had shown that the electrical conductivity of an organic polymer — polyacetylene — could also be increased by a factor of 10^{12} (to ~ 10^3 S/cm) when it was doped with an electron donor such as an alkali-metal ion or an electron acceptor such as arsenic pentafluoride (AsF_5) or iodine. The conductivity of doped polyacetylene is comparable to that of copper on an equal weight basis.

More recently, conductivity has been demonstrated for doped versions of poly(*p*-phenylene), polypyrrole, polythiophene, and polyaniline. Although the electrical conductivities of these polymers are lower than polyacetylene, these polymers are more stable to effects of oxygen and moisture. As shown by the chemical structure of the repeating units of these polymers in Table 4.9, a common feature that makes these polymers capable of transporting electrical charge is a conjugated π-electron consisting of alternation single and double bonds along the polymer chain backbone or ring structure. It is believed that doping results in an electron imbalance, and the extended π-conjugated structure of the conductive polymer allows the new electron population to move along the backbone when an electric potential is applied.

In general, conductivity increases with decreasing band gap which is the amount or energy needed to promote an electron from the highest occupied energy level or valence band to the empty band immediately above it. The empty band is the conduction band. Metals have zero band gaps, while insulators like many polymers have large band-gaps (1.5 to 4 eV) which impede electron flow. By careful design of the chemical structure of the polymer backbone, it may be possible to obtain band gaps as low as 0.5 to 1 eV. Currently, the lowest band gap observed for a polymer is ~ 1 eV reported for polyisothianaphthalene (see Table 4.9).

Table 4.9 Chemical structures and conductivities of some electrically conductive polymers

Conductive polymer	Repeating unit	Dopants	Conductivity[a] (S/cm)
Polyaniline[b]	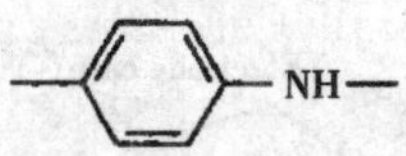	HCl	200
trans-Polyacetylene		I_2, Br_2, Li, Na, AsF_5	10^4
Polypyrrole	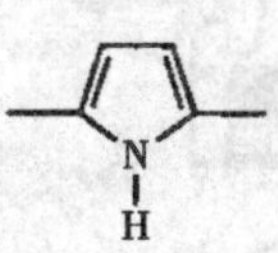	BF_4^-, ClO_4^-, tosylate[c]	500 — 7500
Polyisothianaphthalene	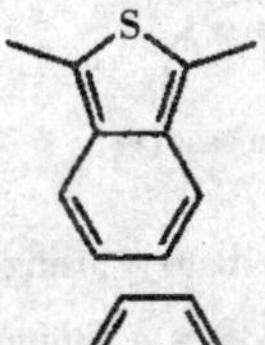	BF_4^-, ClO_4^-	50
Poly(*p* - phenylene)	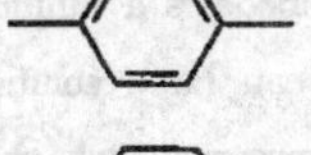	AsF_5, Li, K	10^3
Poly(*p* - phenylene-vinylene)	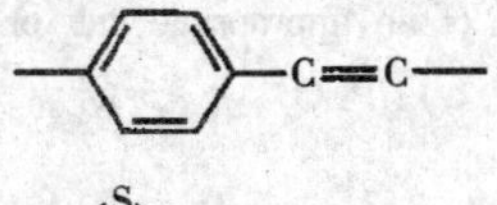	AsF_5	10^4
Polythiophene	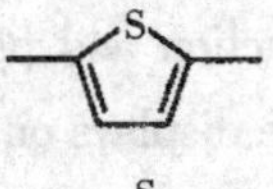	BF_4^-, ClO_4^-, tosylate[c] $FeCl_4^-$	10^3
Poly(3 - alkyl-thiophene)	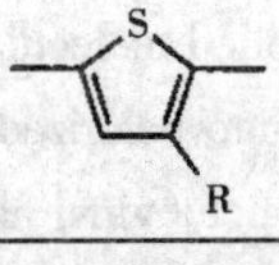	[illegible], ClO_4^-, $FeCl_4^-$,	10^3— 10^4

[a] Approximate maximum conductivity of doped polymer.

[b] Polyaniline exists in four oxidation states, of which only the emeraldine salt

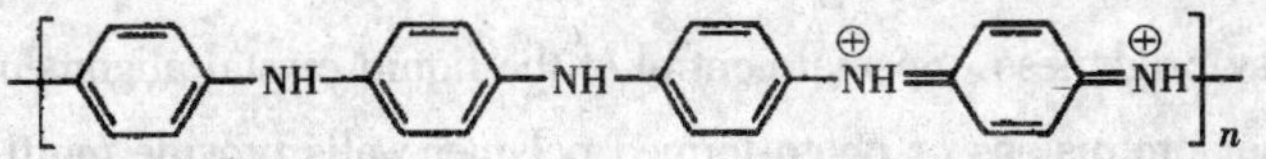

[c] *p* - Methylphenylsulfonate.

Potential application of conductive polymers are significant since they are general lighter, more flexible, and easier to fabricate than many of the materials they seek to replace. Current examples include the use of polymers as both cathodes and solid electrolytes in batteries for automotive and other applications as alternatives to lead-acid batteries in automotive and other applications. A diagram of a typical lithium-polymer battery is shown in Figure 4.24. Potential advantages of polymeric batteries include high reliability, light weight, nonleakage of electrolyte solution, ultrathin-film form, flexibility, and high energy density (up can be used as the cathode material, while lithium dissolved in aluminum foil serves as the anode. Lithium-poly(*p*-phenylene) batteries can deliver current densities as high as 50 mA/cm^2 have efficiencies of 91% in charging and discharging, and have theoretical energy densities of 320 W·h/kg. Examples of polymers used in the formulation of solid electrolytes include poly(alkyl sulfide)s, polyphosphazene having oligo (oxyethylene) side groups, and especially poly(ethylene oxide). Salts that are added to form the electrolyte include a number of lithium salts having low crystal lattice energies, such as lithium tetrafluoroborate ($LiBF_4$), lithium hexafluoroarsenate ($LiAsF_6$), lithium perchlorate ($LiClO_4$), and lithium trifluoromethanesulfonate sometimes called lithium triflate ($LiCF_3SO_3$).

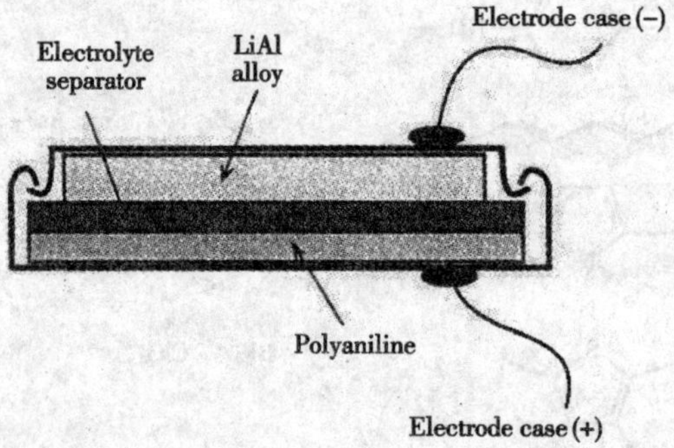

Figure 4.24 Illustration of a rechargeable battery using a polymeric cathode (polyaniline). The anode is a lithium-aluminum alloy, while the electrolyte can be a solution (e.g., propylene carbonate) of an appropriate salt or a solid polymer electrolyte (e.g., Li-PEO)

Polymeric materials play an important role in the construction and performance of liquid crystal displays. New functional polymers are developed to improve the displays on brightness, power efficiency and viewing angle performance. Polymer films with a controlled molecular structure and architecture provide new polarization optics to convert non-polarized light into polarized light with higher efficiency than the traditional polarizers based on stretched iodine doped poly(vinyl alcohol). Other films provide a polarization compensation function to maintain an angular invariant net optical retardation of the display device and therewith a better contrast to wider viewing angles. Color generation by non-absorbing methods is believed to improve on the display brightness. Special control of the liquid crystal alignment by photo-sensitive orientation layers, polymer protrusions or photo-formed polymer walls provide multiple director patterns within a single pixel to average out angular LC effects.

Keywords

amorphous	artificial organ	absorbable and nonabsorbable
biomaterial	cellulose	chitosan
crystallinity	collagen	configuration
controlled drug delivery	conductive polymer	esterification reaction
elastomer	engineering plastic	extrusion
functional group	flexibility	homochain polymer
heterochain polymer	hydrogen-bonding	high-density polyethylene
high-impact polystyrene	kidney dialysis	low-density polyethylene
lithium-polymer battery	liquid crystal display	metallocene catalyst
morphology	microfiltration	natural polymer
nylon	nanofiltration	orthopedic implant
polyethylene	polyesters	polyether
polyurethane	polysaccharide	polyolefin
polypropylene	processability	poly(vinylidene chloride)
plasticizer	poly(ethylene terephthalate)	polycarbonate
polysiloxane	poly(methyl methacrylate)	polymeric membrane
phase inversion	polyglycolide	poly(ethylene glycol)
polyacetylene	polypyrrole	polyaniline
reverse osmosis	repeating unit	semicrystalline material
solvent evaporation	suture	thermoplastic polymer
thermosetting polymer	tissue engineering scaffold	tissue implant
ultrafiltration	vulcanization	viscosity

Problems

1. What is a natural polymer? (A giant molecule found in nature)

2. Draw the monomer structure of polyethylene. ($H_2C{=}CH_2$)

3. What is the repeating linkage in a protein? (the peptide linkage)

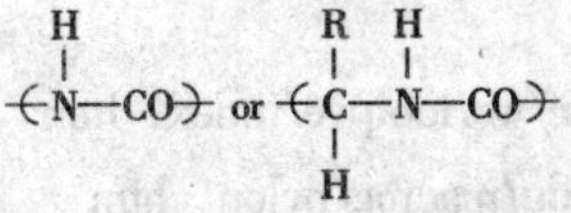

4. Are most commercial plastics thermoplastics or thermosets? (Thermoplastic)

5. Cite the differences in behavior and molecular structure of thermoplastic and thermosetting polymers.

6. Is a crosslink a covalent bond or a hydrogen bond? (Covalent bond)

7. Which is more highly crystalline, HDPE or LDPE? Which has the larger specific volume, HDPE or LDPE? (HDPE, LDPE)

8. Which will have the higher melting point, atactic or isotactic PP? (Isotactic PP is a solid; atactic PP is a soft gummy substance.)

9. Which would be more likely to crystallize, atactic PP or a block copolymer of ethylene and propylene? (The block copolymer providing the sequences were linear and ordered.)

10. Which of the following are engineering resins: polystyrene, nylon, and polycarbonate? (Nylon and polycarbonate.)

11. Which has the higher melting point, nylon 66 or Kevlar? (Kevlar, it is an aromatic nylon.)

12. What is the advantage of a clear polycarbonate sheet over a sheet of poly(methyl methacrylate)? (Polycarbonate is tougher.)

13. What is the principal difference in the structure of a quartz or silica sand and a siloxane polymer (silicone)? (They have similar backbones, but the siloxane polymer has organic pendant groups on the silicon atoms.)

14. What is kidney dialysis? (Application of the polymeric membrane to remove toxins of low to moderate molecular weight from blood)

15. What are categories of suture? (Absorbable and nonabsorbable, monofilament, multifilament, and braided)

16. What is the purpose of controlled drug delivery? (Maintain a steady therapeutic concentration.)

17. What is the special chemical structure of conductive polymers? (A conjugated π-electron consisting of alternation single and double bonds along the polymer chain backbone or ring structure.)

Selected from

[1] P. Bahadur, N. V. Sastry. Principles of Polymer Science. Alpha Science Interational, 2002.

[2] W. D. Callister. Fundamentals of Materials Science and Engineering. John Wiley & Sons, 2001.

[3] M. N. V. Ravi Kumar. Reactive and Functional Polymers. 2000, 46: 1 – 27.

[4] C. E. Carraher. Giant Molecules: Essential Materials for Everyday Living and Problem Solving. 2nd Edition. John Wiley & Sons, 2003.

[5] http: //www. psrc. usm. edu/macrog/pe. htm

[6] J. Scheirs, D. B. Priddy. Modern Styrenic Polymers-Polystyrenes and Styrenic Copolymers. John Wiley & Sons, 2003.

[7] http: //www. evc – int. com/worldofpve/index. htm

[8] http: //www. psrc. usm. edu/macrog/nylon. htm

[9] http: //www. chemheritage. org/EducationalServices/chemach/popwhc/html

[10] http: //www. zimmer – ag. de

[11] http: //www. tamms. com

[12] http: //www. resins. com

[13] J. R. Fried. Polymer Science and Technolgy (聚合物科学与工程). 北京:化学工业出版社, 2005.

[14] S. P. Nunes, K. V. Peinemann. Membrane Technology in the Chemical Industry. Wiley-VCH Verlag GmbH, 2001.

[15] J. B. Park, Roderic S. Lakes. Biomaterials: An Introduction. 2 Edition. New York and London: Plenum Press, 1992.

[16] C. C. Chu, J. A. von Fraunhofer, H. P. Greisler. Wound Closure Biomaterials and Devices. New York: CRC Press, 1997.

5 Composite Materials

5.1 Introduction

Composites are produced when two materials are joined to give a combination of properties that cannot be attained in the original materials. Composite materials may be selected to give unusual combination of stiffness, strength, weight, high-temperature performance, corrosion resistance, hardness, or conductivity.

Composites can be placed into three categories — particulate, fiber, and laminar — based on the shapes of the materials (Figure 5.1). Concrete, a mixture of cement and gravel, is a particulate composite; fiberglass, containing glass fibers embedded in a polymer, is a fiber-reinforced composite; and plywood, having alternating layers of wood veneer, is a laminar composite. If the reinforcing particles are uniformly distributed, particulate composites have isotropic properties; fiber composites may be either isotropic or anisotropic; laminar composites always display anisotropic behavior.

(a) (b) (c)

Figure 5.1 Some examples of composite materials: (a) plywood is a laminar composite of layers of wood veneer, (b) fiberglass is a fiber-reinforced composite containing stiff, strong glass fibers in a softer polymer matrix ($\times$ 175), (c) concrete is a particulate composite containing coarse sand or gravel in a cement matrix (reduced 50%)

5.2 Dispersion-strengthened composites

By stretching slightly our definition of a composite, we can consider a special group of dispersion-strengthened materials containing particulates 10 nm to 250 nm in diameter as particulate composites. These dispersoids, usually a metallic oxide, are introduced into the matrix by means other than traditional phase transformations. Even though the small particles are not coherent with the matrix, they block the movement of dislocations and produce a pronounced strengthening effect.

At room temperature, the dispersion-strengthened composites may be weaker than traditional age-hardened alloys, which contain a coherent precipitate. However, because the composites do not catastrophically soften by overaging, overtempering, grain growth, or coarsening of the dispersed phase, the strength of the composite decreases only gradually with increasing temperature (Figure 5.2). Furthermore, their creep resistance is superior to that of metals and alloys.

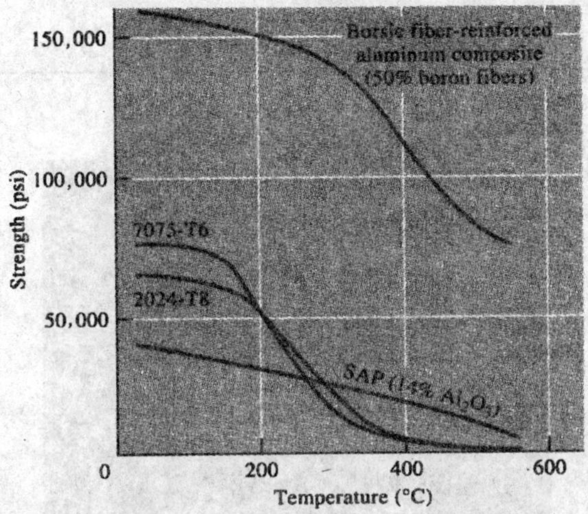

Figure 5.2 Comparison of the yield strength of dispersion-strengthened sintered aluminum powder (SAP) composite with that of two conventional two-phase high-strength aluminum alloys. The composite has benefits above about 300 ℃. A fiber-reinforced aluminum composite is shown for comparison

The dispersant must have a low solubility in the matrix and must not chemically react with the matrix, but a small amount of solubility may help improve the bonding between the dispersant and the matrix. Copper oxide (Cu_2O) dissolves in copper at high temperatures; thus, the Cu_2O-Cu system would not be effective. However, Al_2O_3 does not dissolve in aluminum; the Al_2O_3-Al system does give an effective dispersion-strengthened material.

Examples of dispersion-strengthened composites

Table 5.1 lists some materials of interest. Perhaps the classic example is the sintered aluminum powder (SAP) composite. SAP has an aluminum matrix strengthened by up to 14% Al_2O_3. The composite is formed by powder metallurgy. In one method, aluminum and aluminum powders are blended, compacted at high pressures, and sintered. In a second technique, the aluminum powder is treated to add a continuous oxide film on each particle. When the powder is compacted, the oxide film fractures into tiny flakes that are surrounded by the aluminum metal during sintering.

Another important group of dispersion-strengthened composites includes thoria-dispersed metals such as TD-nickel (Figure 5.3). TD-nickel can be produced by internal oxidation. Thorium is present in nickel as an alloying element. After a powder metallurgy compact is made, oxygen is allowed to diffuse into the metal, react with the thorium, and produce thoria (ThO_2).

Table 5.1 Examples and applications of selected dispersion-strengthened composites

System	Applications
Ag-CdO	Electrical contact materials
$Al-Al_2O_3$	Possible use in nuclear reactors
Be-BeO	Aerospace and nuclear reactors
$Co-ThO_2$, Y_2O_3	Possible creep-resistant magnetic materials
Ni-20% $Cr-ThO_2$	Turbine engine components
Pb-PbO	Battery grids
$Pt-ThO_2$	Filaments, electrical components
$W-ThO_2$, ZrO_2	Filaments, heaters

Figure 5.3 Electron micrograph of TD-nickel. The dispersed ThO_2 particles have a diameter of 300 nm or less (× 2,000)

5.3 True particulate composites

The true particulate composites contain large amount of coarse particles that do not effectively block slip. The particulate composites are designed to produce unusual combinations of properties rather than to improve strength.

Rule of mixtures

Certain properties of a particulate composite depend only on the relative amounts and properties of the individual constituents. The rule of mixtures can accurately predict these properties. The density of a particulate composite, for example, is

$$\rho_c = \Sigma f_i \rho_i = f_1 \rho_1 + f_2 \rho_2 + \cdots + f_n \rho_n \tag{5.1}$$

where ρ_c is the density of the composite, $\rho_1, \rho_2, \ldots, \rho_n$ are the densities of each constituent in the composite, and V_1, V_2, ..., V_n are the volume fractions of each constituent.

Cemented carbides

Cemented carbides, or cermets, contain hard ceramic particles dispersed in a metallic matrix. Tungsten carbide inserts used for cutting tools in machining operations are typical of this group. Tungsten carbide (WC) is a hard, stiff, high-melting-temperature ceramic. Unfortunately, tools constructed from tungsten carbide are extremely brittle.

To improve toughness, tungsten carbide particles are combined with cobalt powder and pressed into powder compacts. The compacts are heated above the melting temperature of the cobalt. The liquid cobalt surrounds each of the solid tungsten carbide particles (Figure 5.4). After solidification, the cobalt serves as the binder for tungsten carbide and provides good impact resistance. Other carbides, such as TaC and TiC, may also be included in the cement.

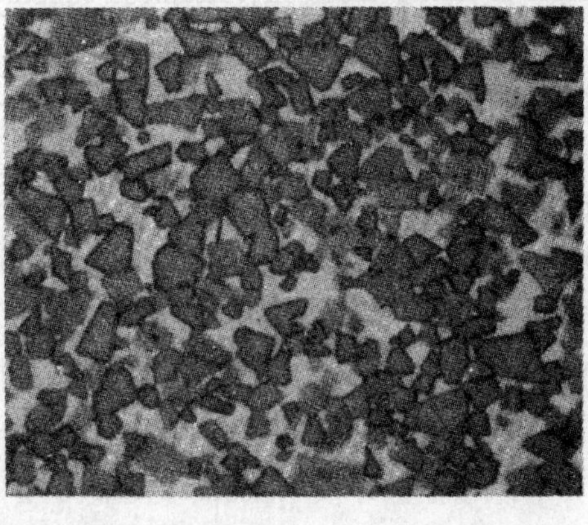

Figure 5.4 Microstructure of tungsten carbide-20% cobalt-cemented carbide (×1 300)

Abrasives

Grinding and cutting wheels are formed from aluminum (Al_2O_3), silicon carbide (SiC), and cubic boron nitride (BN). To provide toughness, the abrasive particles are bonded by a glass or polymer matrix. Diamond abrasives are typically bonded with a metal matrix. As the hard particles wear, they fracture or pull out of the matrix, exposing new cutting surfaces.

Electrical contacts

Materials used for electrical contacts in switches and relays must have a good combination of wear resistance and electrical conductivity. Otherwise, the contacts erode, causing poor contact and arcing. Tungsten-reinforced silver provides this combination of characteristics. A tungsten powder compact is made using conventional powder metallurgical process (Figure 5.5) to produce high interconnected porosity. Liquid silver is then vacuum infiltrated to fill the interconnected voids. Both the silver and the tungsten are continuous. Thus, the pure silver efficiently conducts current while the hard tungsten provides wear resistance.

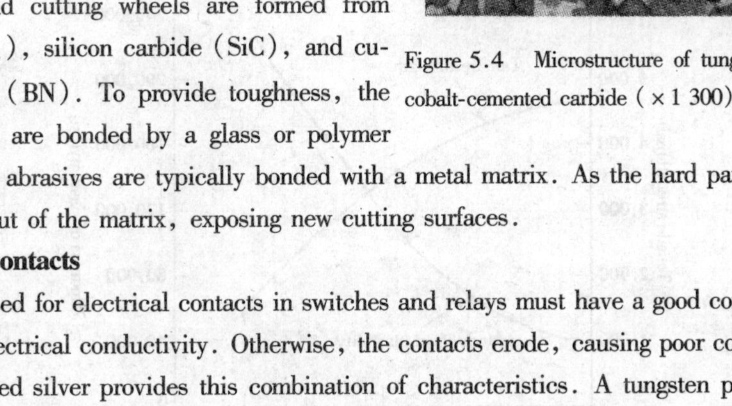

Figure 5.5 The steps in producing a silver-tungsten electrical composite: (a) tungsten powders are pressed, (b) a low-density compact is produced, (c) sintering joins the tungsten powders, and (d) liquid silver is infiltrated into the pores between the particles

Polymers

Many engineering polymers that contain fillers and extenders are particulate composites. A classic example is carbon black in vulcanized rubber. Carbon black consists of tiny carbon spheroids only 5 nm to 500 nm in diameter. The carbon black improves the strength, stiffness, hardness, wear resistance, and heat resistance of the rubber.

Extenders, such as calcium carbonate, solid glass spheres, and various clays, are added so that a smaller amount of the more expensive polymer is required. The extenders may stiffen the polymer, increase the hardness and wear resistance, increase thermal conductivity, or improve resistance to creep; however, strength and ductility normally decrease (Figure 5.6). Introducing hollow glass spheres may impact the same changes in properties while significantly reducing the weight of the composite.

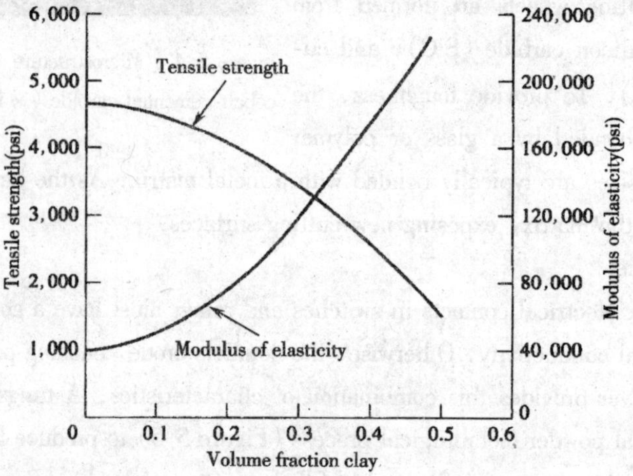

Figure 5.6 The effect of clay on the properties of polyethylene

Other special properties can be obtained. Elastomer particles are introduced into polymer to improve toughness. Polyethylene may contain metallic powders, such as lead, to improve absorption of fission products in nuclear applications.

Cast metal particulate composites

Aluminum castings containing dispersed SiC particles for automotive application, including pistons and connecting rods, represent an important commercial application for particulate composites (Figure 5.7). With special processing, the SiC particles can be wet by the liquid, helping to keep the ceramic particles from sinking during freezing.

An unusual technique for producing particulate-reinforced castings is based on the thixotropic behavior of partly liquid-partly solid melts. A liquid alloy is allowed to cool until about 40% solids have formed; during solidification, the solid-liquid mixture is vigorously stirred to break up the dendritic structure (Figure 5.8). A particulate material is introduced during stirring. The resulting solid-liquid slurry displays thixotropic behavior — the slurry behaves as a solid when no stress is applied, but flows like a liquid when pressure is exerted. Consequently, the thixotropic slurry can be injected into a die under pressure, a process called compocasting. A variety of ceramic particles and glass beads have been incorporated into aluminum and magnesium alloys by this technique.

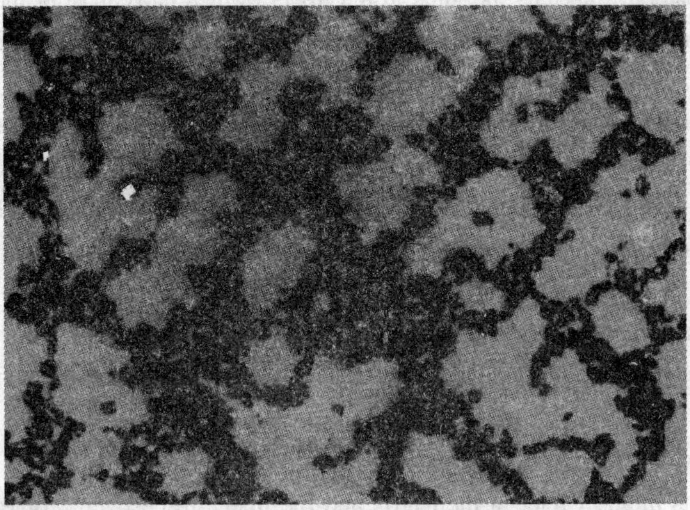

Figure 5.7 Microstructure of an aluminum casting alloy reinforced with silicon carbide particles. In this case, the reinforcing particles have segregated to interdendritic regions of the casting (× 125)

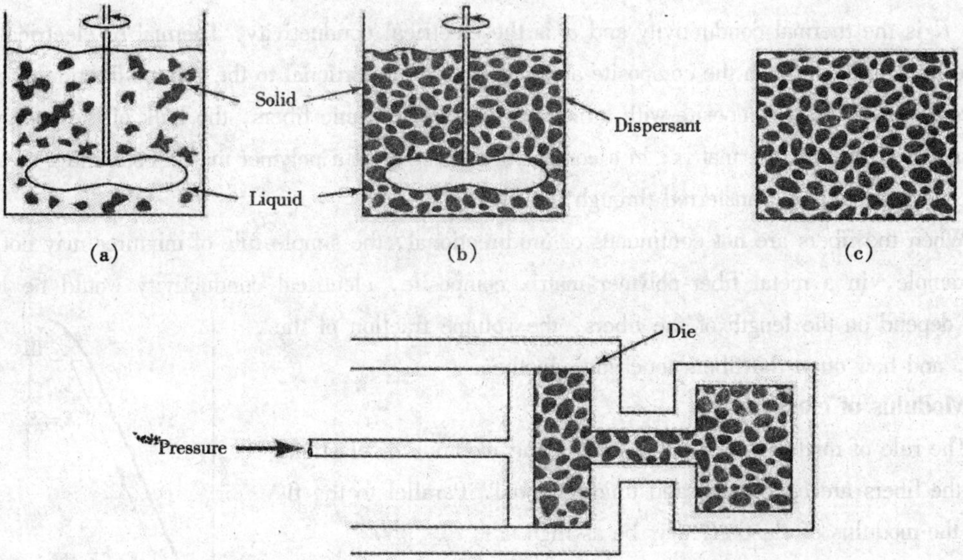

Figure 5.8 In compocasting, (a) a solidifying alloy is stirred to break up the dendritic network, (b) a reinforcement is introduced into the slurry, (c) when no force is applied, the solid-liquid mixture does not flow, and (d) high pressure cause the solid-liquid mixture to flow into a die

5.4 Fiber-reinforced composites

Most fiber-reinforced composites provide improved strength, fatigue resistance, stiffness, and strength-to-weight ratio by incorporating strong, stiff, but brittle fibers into a softer, more ductile matrix. The matrix material transmits the force to the fibers, which carry most of the applied force. The strength of the composite may be high at both room temperature and elevated temperatures (Figure 5.2).

Many types of reinforcing materials are employed. Straw has been used to strengthen mud bricks for centuries. Steel reinforcing bars are introduced into concrete structures. Glass fibers in a polymer matrix produce fiberglass for transportation and aerospace applications. Fibers made of boron, carbon, polymers, and ceramics provide exceptional reinforcement in advanced composites based on materials of polymers, metals, ceramics, and even intermetallic compounds.

The rule of mixtures in fiber-reinforced composites

As for fiber composites, the rule of mixtures always predicts the density of fiber-reinforced composites:

$$\rho_c = f_m \rho_m + f_f \rho_f \tag{5.2}$$

where the subscripts m and f refer to the matrix and the fiber, respectively. Note that $f_m = 1 - f_f$.

In addition, the rule of mixtures accurately predicts the electrical and thermal conductivity of fiber-reinforced composites along the fiber direction if the fibers are continuous and unidirectional:

$$K_c = f_m K_m + f_f K_f \tag{5.3}$$

$$\sigma_c = f_m \sigma_m + f_f \sigma_f \tag{5.4}$$

where K is the thermal conductivity and σ is the electrical conductivity. Thermal or electrical energy can be transferred through the composite at a rate that is proportional to the volume fraction of the conductive material. In a composite with a metal matrix and ceramic fibers, the bulk of the energy would be transferred through the matrix; in a composite consisting of a polymer matrix containing metallic fibers, energy would be transferred through the fibers.

When the fibers are not continuous or unidirectional, the simple rule of mixtures may not apply. For example, in a metal fiber-polymer matrix composite, electrical conductivity would be low and would depend on the length of the fibers, the volume fraction of the fibers, and how often the fibers touch one another.

Modulus of elasticity

The rule of mixtures is used to predict the modulus of elasticity when the fibers are continuous and unidirectional. Parallel to the fibers, the modulus of elasticity may be as high as:

$$E_c = f_m E_m + f_f E_f \tag{5.5}$$

However, when the applied stress is very large, the matrix begins to deform and the stress-strain curve is no longer linear (Figure 5.9). Since the matrix now contributes little to the stiffness of the composite, the modulus can be approximated by:

$$E_c = f_f E_f \tag{5.6}$$

When the load is applied perpendicular to the fibers, each component of the composite acts independently of the other. The modulus of the composites is now:

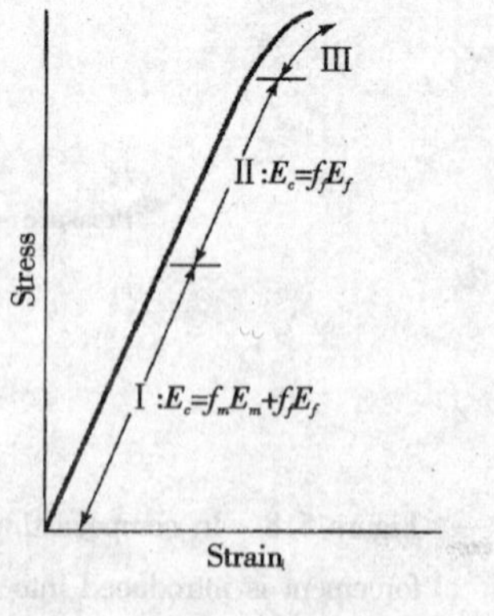

Figure 5.9 The stress-strain curve for a fiber-reinforced composite. At low stresses, the modulus of elasticity is given by the rule of mixtures. At higher stresses, the matrix deforms and the rule of mixtures is no longer obeyed

$$\frac{1}{E_c} = \frac{f_m}{E_m} + \frac{f_f}{E_f} \tag{5.7}$$

Again, if the fibers are not continuous or unidirectional, the rule of mixtures does not apply.

Strength of composites

The strength of a fiber-reinforced composite depends on the bonding between the fibers and the matrix. However, the rule of mixtures is sometimes used to approximate the tensile strength of a composite containing continuous, parallel fibers:

$$\sigma_c = f_f \sigma_f + f_m \sigma_m \tag{5.8}$$

where σ_f is the tensile strength of the fiber and σ_m is the stress acting on the matrix when the composite is strained to the point where the fiber fractures. Thus, σ_m is not the actual tensile strength of the matrix. Other properties, such as ductility, impact properties, fatigue properties, and creep properties, are difficult to predict even for unidirectionally aligned fibers.

5.5 Characteristics of fiber-reinforced composites

Many factors must be considered when designing a fiber-reinforced composite, including the length, diameter, orientation, amount, and properties of the fibers; the properties of the matrix; and the bonding between the fibers and matrix.

Fiber length and diameter

Fiber can be short, long, or even continuous. Their dimensions are often characterized by the aspect ratio l/d, where l is the fiber length and d is the diameter. Typical fibers have diameters varying from 10 microns (10×10^{-4} cm) to 150 microns (150×10^{-4} cm).

The strength of a composite improves when the aspect ratio is large. Fibers often fracture because of surface imperfections. Making the diameter as small as possible gives the fiber less surface imperfections. Making the diameter as small as possible gives the fiber less surface area and, consequently, fewer flaws that might propagate during processing or under load. We also prefer long fibers. The ends of a fiber carry less of the load than the remainder of the fiber; consequently, the fewer the ends, the higher the load-carrying ability of the fibers (Figure 5.10).

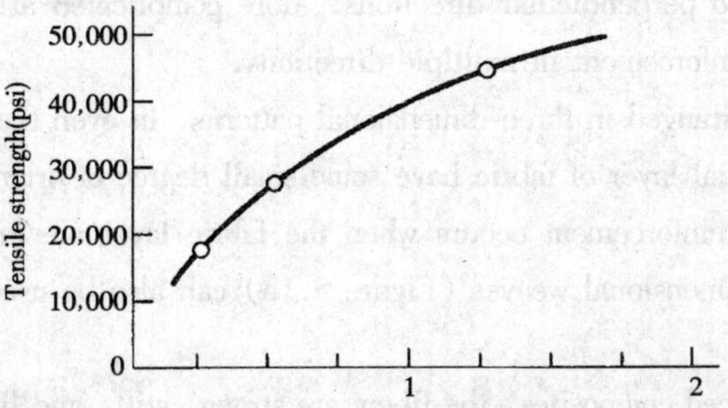

Figure 5.10 Increase the length of chopped E-glass fibers in an epoxy matrix increases the strength of the composite. In this example, the volume fraction of glass fibers is about 0.5

In many fiber-reinforced systems, discontinuous fibers with an aspect ratio greater than some critical value are used to provide an acceptable compromise between processing ease and properties. A critical fiber length l_c, for any given fiber diameter d, can be determined:

$$l_c = \frac{\sigma_f d}{2\tau_i} \tag{5.9}$$

where σ_f is the strength of the fiber and τ_i is related to the strength of the bond between the fiber and the matrix, or the stress at which the matrix begins to deform. If the fiber length l is smaller than l_c, little reinforcing effect is observed; if l is greater than about 15 l_c, the fiber behaves almost as if it were continuous. The strength of the composite can be estimated from

$$\sigma_c = f_f \sigma_f \left(1 - \frac{l_c}{2l}\right) + f_m \sigma_m \tag{5.10}$$

where σ_m is the stress on the matrix when the fibers break.

Amount of fibers

A greater volume fraction of fibers increases the strength and stiffness of the composite, as we would expect from the rule of mixtures. However, the maximum volume fraction is about 80%, beyond which fibers can no longer be completely surrounded by the matrix.

Orientation of fibers

The reinforcing fibers may be introduced into the matrix in a number of orientations. Short, randomly oriented fibers having a small aspect ratio — typical of fiberglass — are easily introduced into the matrix and give relatively isotropic behavior in the composite.

Long, or even continuous, unidirectional arrangements of fibers produce anisotropic properties, with particularly good strength and stiffness parallel to the fibers. These fibers are often designated as 0° plies, indicating that all of the fibers are aligned with the direction of the applied stress. However, unidirectional orientations provide poor properties if the load is perpendicular to the fibers (Figure 5.11).

One of the unique characteristics of fiber-reinforced composites is that their properties can be tailored to meet different types of loading conditions. Long, continuous fibers can be introduced in several directions within the matrix (Figure 5.12); in orthogonal arrangements (0°/90° plies), good strength is obtained in two perpendicular directions. More complicated arrangements (such as 0°/±45°/90° plies) provide reinforcement in multiple directions.

Fibers can also be arranged in three-dimensional patterns. In even the simplest of fabric weaves, the fibers in each individual layer of fabric have some small degree of orientation in a third direction. Better three-dimensional reinforcement occurs when the fabric layer are knitted or stitched together. More complicated three-dimensional weaves (Figure 5.13) can also be used.

Fiber properties

In most fiber-reinforced composites, the fibers are strong, stiff, and lightweight. If the composite is to be used at elevated temperatures, the fiber should also have a high melting temperature. Thus the specific strength and specific modulus of the fiber are important characteristics:

$$\text{Specific strength} = \frac{\sigma_y}{\rho} \tag{5.11}$$

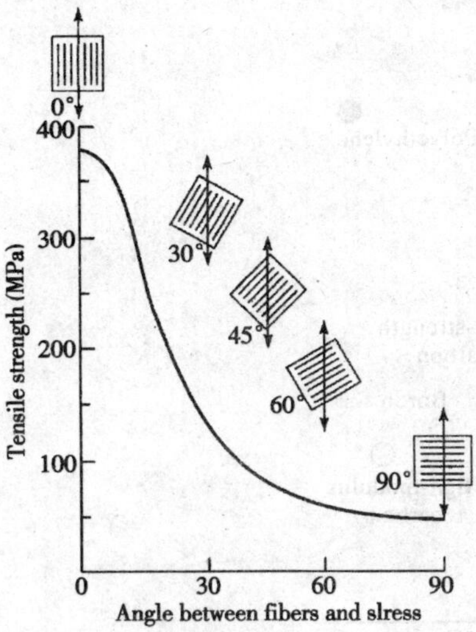

Figure 5.11 Effect of fiber orientation on the tensile strength of E-glass fiber-reinforced epoxy composites

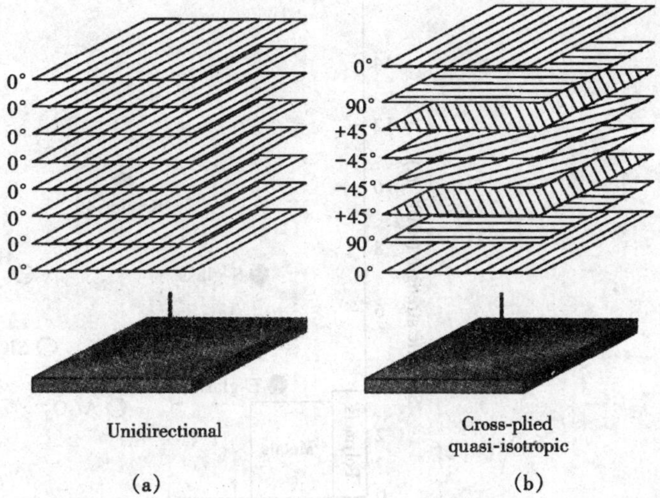

Figure 5.12 (a) Tapes containing aligned fibers can be joined to produce a multi-layered unidirectional composite structure. (b) Tapes containing aligned fibers can be joined with different orientations to produce a quasi-isotropic composite. In this case, a 0°/ ± 45°/90° composite is formed

Figure 5.13 A three-dimensional weave for fiber-reinforced composites

$$\text{Specific modulus} = \frac{E}{\rho} \tag{5.12}$$

where σ_y is the yield strength, ρ is the density, and E is the modulus of elasticity. Properties of typical fibers are shown in Figure 5.14. The highest specific modulus is usually found in materials having a low atomic number and covalent bonding, such as carbon and boron. These two elements also have a high strength and melting temperature.

Aramid fibers, of which Kevlar is the best known example, are aromatic polyamide polymers strengthened by a backbone containing benzene rings (Figure 5.15) and are examples of liquid-crystalline polymers in that the polymer chain are rodlike and very stiff. Specially prepared polyethylene fibers are also available. Both the aramid and polyethylene fibers have excellent strength and stiffness but are limited to low-temperature use. Because of their lower density, polyethylene fibers have superi-

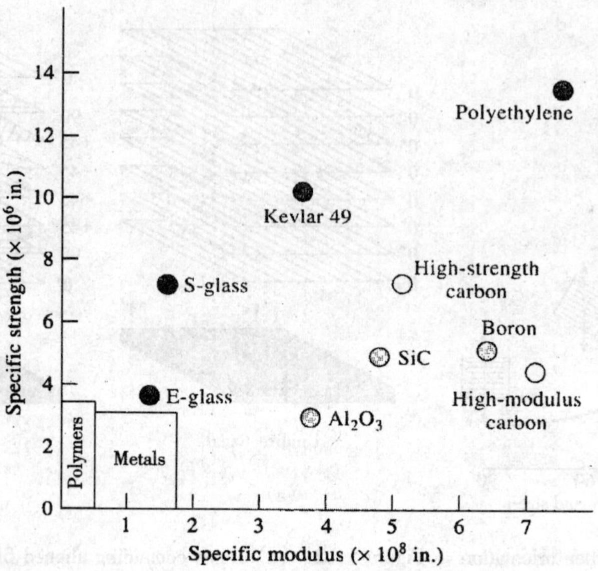

Figure 5.14 Comparison of the specific strength and specific modulus of fibers versus metals and polymers

or specific strength and specific modulus.

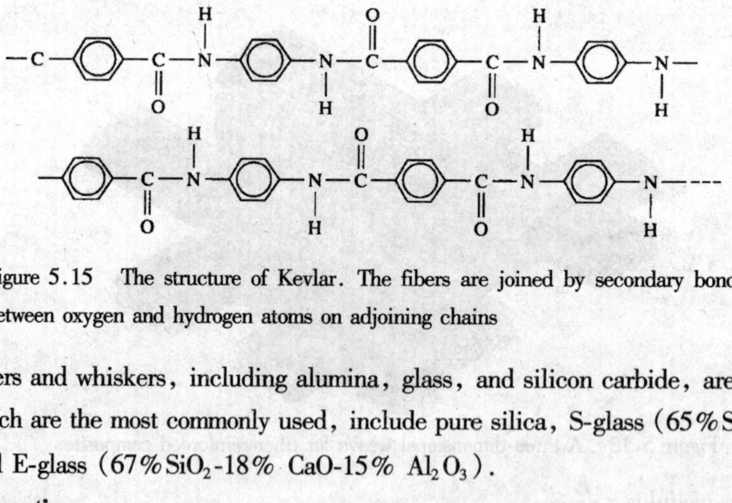

Figure 5.15 The structure of Kevlar. The fibers are joined by secondary bonds between oxygen and hydrogen atoms on adjoining chains

Ceramic fibers and whiskers, including alumina, glass, and silicon carbide, are strong and stiff. Glass fibers, which are the most commonly used, include pure silica, S-glass (65% SiO_2-25% Al_2O_3-10% MgO), and E-glass (67% SiO_2-18% CaO-15% Al_2O_3).

Matrix properties

The matrix supports the fibers and keeps them in the proper position, transfers the load to the strong fibers, protects the fibers from damage during manufacture and use of the composite, and prevents cracks in the fiber from propagating throughout the entire composite. The matrix usually provides the major control over electrical properties, chemical behavior, and elevated-temperature use of the composite.

Polymer matrices are particularly common. Most polymer materials - both thermoplastics and thermosets - are available in short glass fiber-reinforced grades. Sheet-molding compounds (SMCs) and bulk-molding compounds (BMCs) are typical of this type of composite. Thermosetting aromatic polyimides are used for somewhat higher temperature applications.

Metal matrix composites include aluminum, magnesium, copper, nickel, and intermetallic compound alloys reinforced with ceramic and metal fibers. A variety of aerospace and automotive applications are satisfied by the MMCs. The metal matrix permits the composite to operate at high temperatures, but producing the composite is often more difficult and expensive than producing the polymer matrix materials.

Amazingly, brittle ceramics may be used as a matrix in composites. The ceramic matrix composites have good properties at elevated temperatures and are lighter in weight than the high-temperature metal matrix composites. In a later section, we will discuss how to develop toughness in CMCs.

Bonding and failure

Particularly in polymer and metal matrix composites, good bonding must be obtained between the various constituents. The fibers must be firmly bonded to the matrix materials if the load is to be properly transmitted from the matrix to the fibers. In addition, the fibers may pull out of the matrix during loading, reducing the strength and fracture resistance of the composite, if bonding is poor. Figure 5.16 illustrates poor bonding of carbon fibers in a copper matrix. In some cases, special coatings may be used to improve bonding. Glass fibers may be coated with a silane "keying" agent (called sizing) to improve bonding and moisture resistance in fiberglass composites. Carbon fibers are similarly coated with an organic material to improve bonding. Boron fibers have been coated with silicon carbide or boron nitride to improve bonding with an aluminum matrix; in fact, these fibers have been called Borsic fibers to reflect the presence of the silicon carbide (SiC) coating.

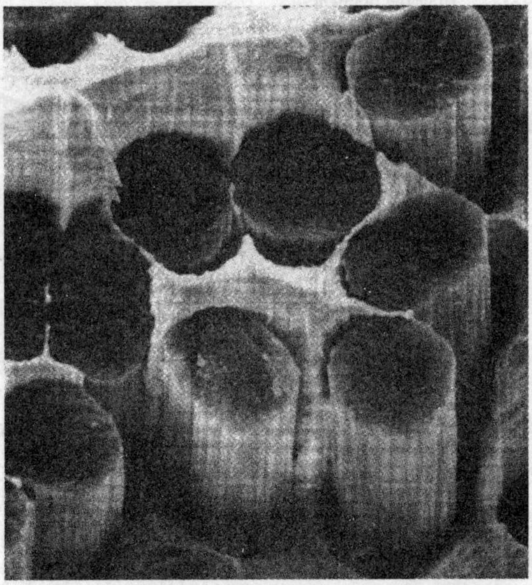

Figure 5.16 Scanning electron micrograph of the fracture surface of a silver-copper alloy reinforced with carbon fibers. Poor bonding causes much of the fracture surface to follow the interface between the metal matrix and the carbon tows (×3,000)

Another property that must be considered when combining fibers into a matrix is the similarity between the coefficients of thermal expansion for the two materials. If the fiber expands and contracts at a rate much different from that of the matrix, fibers may break or bonding can be disrupted, causing premature failure.

In many composites, individual plies or layers of fabric are joined. Bonding between these layers must also be good or another problem, delamination, may occur. The layers may tear apart under load and cause failure. Using composites with a three-dimensional weave will help prevent delamination.

5.6 Manufacturing fibers and composites

Producing a fiber-reinforced composite involves several steps, including producing the fibers, arranging the fibers into bundles or fabrics, and introducing the fibers into the matrix.

Making the fibers

Metallic fibers, glass fibers, and many polymer fibers including nylon, aramid, and polyacrylonitrile (PAN) can be formed by drawing processes.

Boron, carbon, and ceramics are too brittle and reactive to be worked by conventional drawing processes. Boron fiber is produced by chemical vapor deposition (CVD) (Figure 5.17 (a)). A very fine heated tungsten filament is used as a substrate, passing through a seal into a heated chamber, decompose, and permit boron to precipitate onto the tungsten wire (Figure 5.18). SiC fibers are made in a similar manner, with carbon fibers as the substrate for the vapor deposition of silicon carbide.

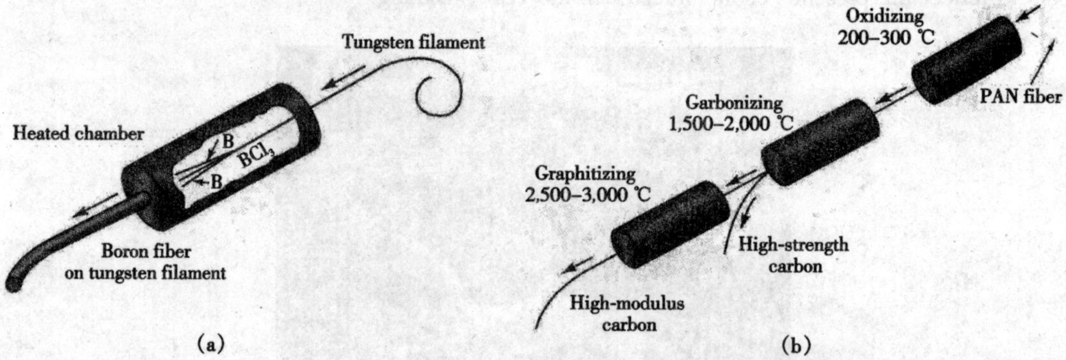

Figure 5.17 Methods for producing (a) boron and (b) carbon fibers

Carbon fibers are made by carbonizing, or pyrolizing, an organic filament, which is more easily drawn or spun into thin, continuous lengths (Figure 5.17(b)). The organic filament, known as a precursor, is often rayon (a cellulosic polymer), polyacrylonitrile, or pitch (various aromatic organic compounds). High temperatures decompose the organic polymer, burning off all of the elements but carbon. As the carbonizing temperature increases from 1,000 ℃ to 3,000 ℃, the tensile strength decreases while the modulus of elasticity increases (Figure 5.19). Drawing the carbon filaments at critical times during carbonizing may produce desirable preferred orientation in the final carbon filament.

Whiskers are single crystals with aspect ratios of 20 to 1,000. Because the whiskers contain no mobile dislocations, slip cannot occur and they have exceptionally high strengths.

Arranging the fibers

Exceptionally fine filaments are bundled together as rovings, yarns, or tows. In yarns, as many

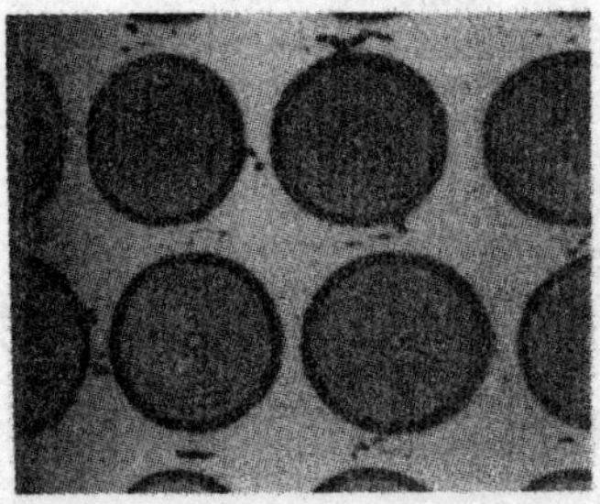

(a)

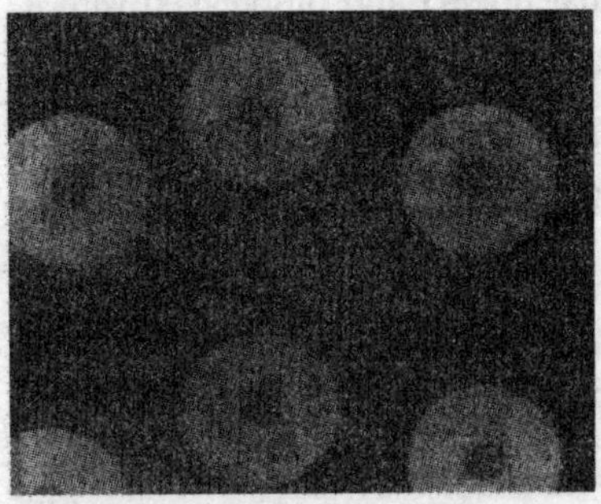

(b)

Figure 5.18 Photomicrographs of two fiber-reinforced composites: (a) the fibers are composed of a thick layer of boron deposited on a small-diameter tungsten filament (b) the fibers are composed of tungsten core and precipitated boron

as 10,000 filaments are twisted together to produce the fiber. A tow contains a few hundred to more than 100,000 untwisted filaments (Figure 5.20). Rovings are untwisted bundles of filaments, yarns, or tows.

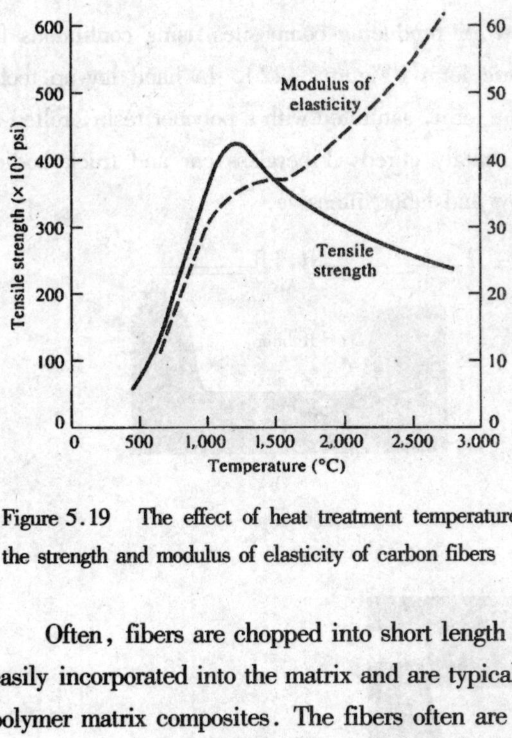

Figure 5.19 The effect of heat treatment temperature on the strength and modulus of elasticity of carbon fibers

Figure 5.20 A scanning electron micrograph of a carbon tow containing many individual carbon filaments (×200)

Often, fibers are chopped into short length of 1 cm or less. These fibers, also called staples, are easily incorporated into the matrix and are typical of the sheet molding and bulk molding compounds for polymer matrix composites. The fibers often are present in the composite in a random orientation.

Long or continuous fibers for polymer matrix composites can be processed into mats or fabrics. Mats contain nonwoven, randomly oriented fibers loosely held together by a polymer resin. The fibers can also be woven, braided, or knitted into two-dimensional or three-dimensional fabrics that are then impregnated with a polymer resin. The resins at this point in the processing have not yet been completely polymerized; these mats or fabrics are called prepregs.

When unidirectionally aligned fibers are to be introduced into a polymer matrix, tapes may be produced. Individual fibers can be unwound from spools onto a mandrel, which determines the spacing of the individual fibers, and prepregged with a polymer resin. These tapes, only one fiber diameter thick, may be up to 48 in. wide. Figure 5.21 illustrates that tapes can also be produced by covering the fibers with upper and lower layers of metal foil that are then joined by diffusion bonding.

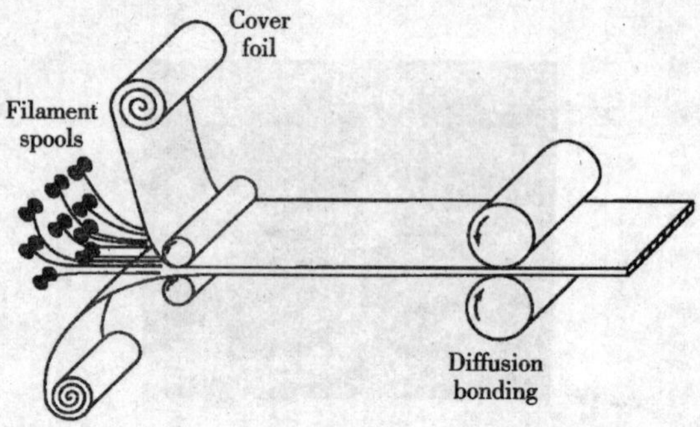

Figure 5.21 Production of fiber tapes by encasing fibers between metal cover sheets by diffusion bonding

Producing the composites

A variety of methods for producing composite parts are used, depending on the application and materials. Short-fiber-reinforced composites are normally formed by mixing the fibers with a liquid or plastic matrix, then using relatively conventional techniques such as injection molding for polymer matrix composites or casting for metal matrix composites. Polymer matrix composites can also be produced by a spray-up method, in which short fibers mixed with a resin are sprayed against a form and cured.

Special techniques, however, have been devised for producing composites using continuous fibers, either in unidirectionally aligned, mat, or fabric form (Figure 5.22). In hand lay-up techniques, the tapes, mats, or fabrics are placed against a form, saturated with a polymer resin, rolled to assure good contact and freedom from porosity, and finally cured. Fiberglass car and truck bodies might be made in this manner, which is generally slow and labor intensive.

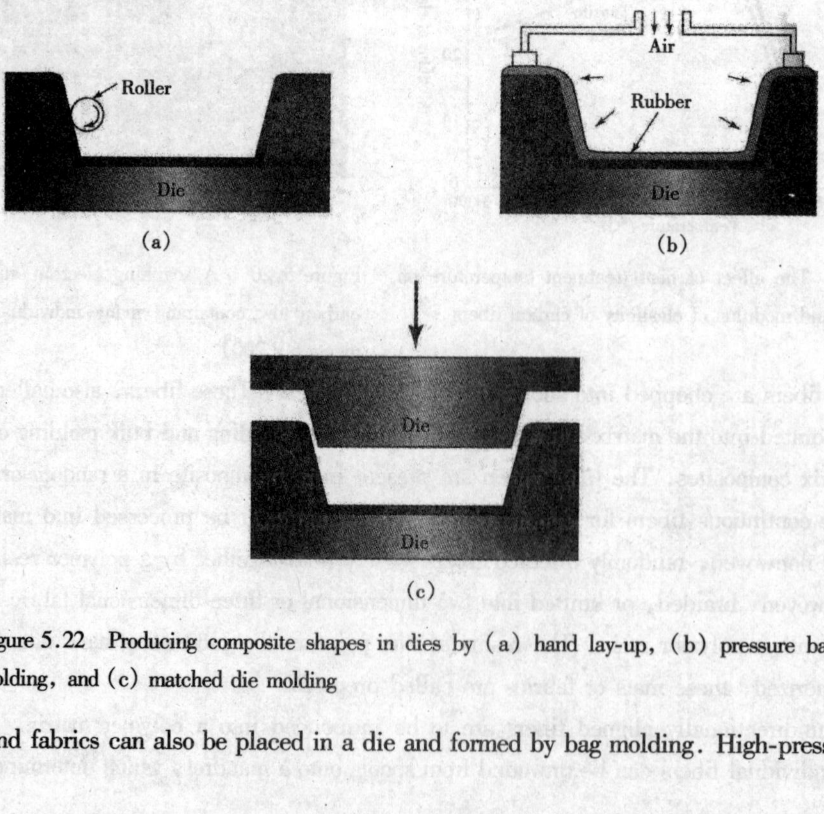

Figure 5.22 Producing composite shapes in dies by (a) hand lay-up, (b) pressure bag molding, and (c) matched die molding

Tapes and fabrics can also be placed in a die and formed by bag molding. High-pressure gases or a vacuum are introduced to force the individual plies together so that good bonding is achieved during curing. Large polymer matrix components for the skins of military aircraft have been produced by these techniques. In matched die molding, short fibers or mats are placed into a two-part die; when the die is closed, the composite shape is formed.

Filament winding is used to produce products such as pressure tanks and rocket motor castings (Figure 5.23). Fibers are wrapped around a form or mandrel to gradually build up a solid shape that may be even several feet in thickness. The filament can be dipped in the polymer matrix resin prior to winding, or the resin can be impregnated around the fiber during or after winding.

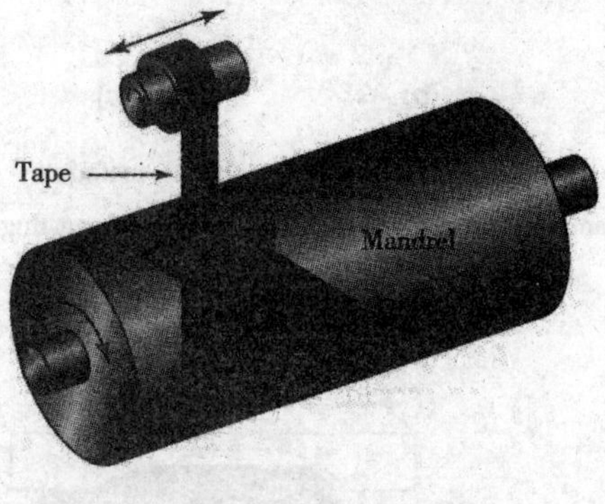

Figure 5.23 Producing composite shapes by filament winding

Pultrusion is used to form a simple-shaped product with a constant cross section, such as round, rectangular, pipe, plate, or sheet shapes (Figure 5.24).

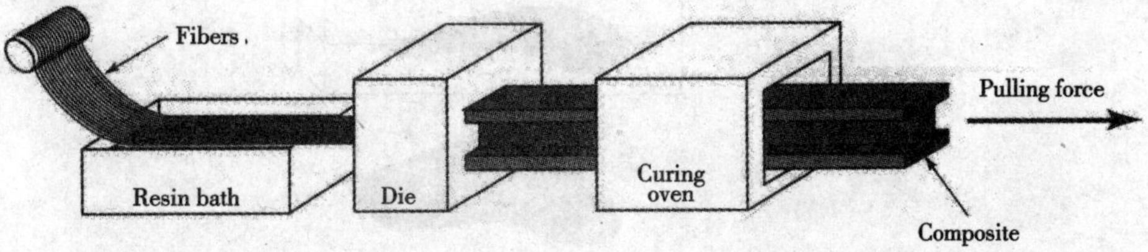

Figure 5.24 Producing composite shapes by pultrusion

Fibers or mats are drawn from spools, passed through a polymer resin bath for impregnation, and gathered together to produce a particular shape before entering a heated die for curing. Curing of the resin is accomplished almost immediately, so a continuous product is produced. The pultruded stock can subsequently be formed into somewhat more complicated shapes, such as fishing poles, golf club shafts, and ski poles.

Metal matrix composites with continuous fibers are more difficult to produce than the polymer matrix composites. Casting processes that force liquid around the fibers using capillary rise, pressure casting, vacuum infiltration, or continuous casting are illustrated in Figure 5.25. Various solid-state compaction processes can also be used. Figure 5.26 illustrates how several tapes can be placed into a closed die and deformed into shape; interdiffusion between the individual tapes produces a solid form.

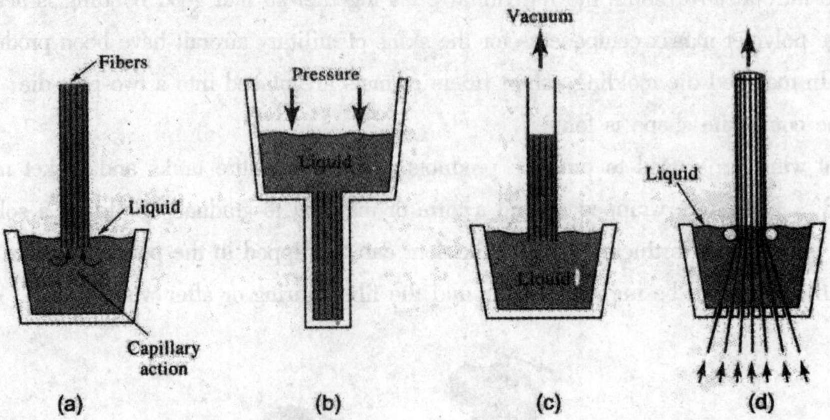

Figure 5.25 Casting techniques for producing composite materials: (a) capillary rise, (b) pressure casting, (c) vacuum infiltration, and (d) continuous casting

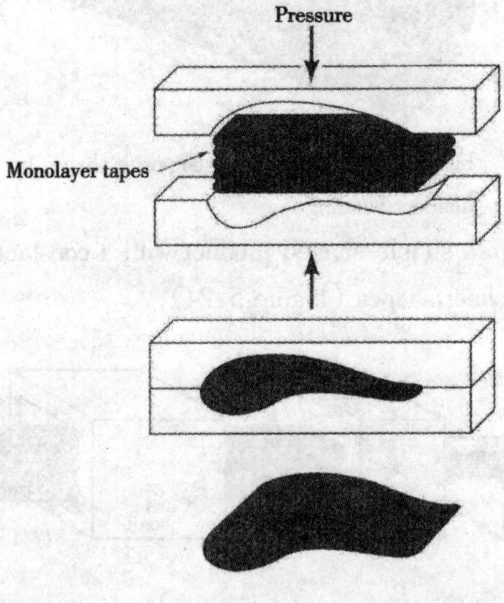

Figure 5.26 Closed die deformation and bonding of a composite composed of several tapes

5.7 Fiber-reinforced systems and applications

Before completing our discussion of fiber-reinforced composites, let's look at the behavior and applications of some the most common of these materials. Figure 5.27 compares the specific modulus and specific strength of several composites with those of metals and polymers. Note that the values in this figure are lower than those in Figure 5.14, since we are now looking at the composite, not just the fiber.

Advanced composites

The term advanced composites is often used when the composite is intended to provide service in very critical applications, as in the aerospace industry (Table 5.2). The advanced composites normal-

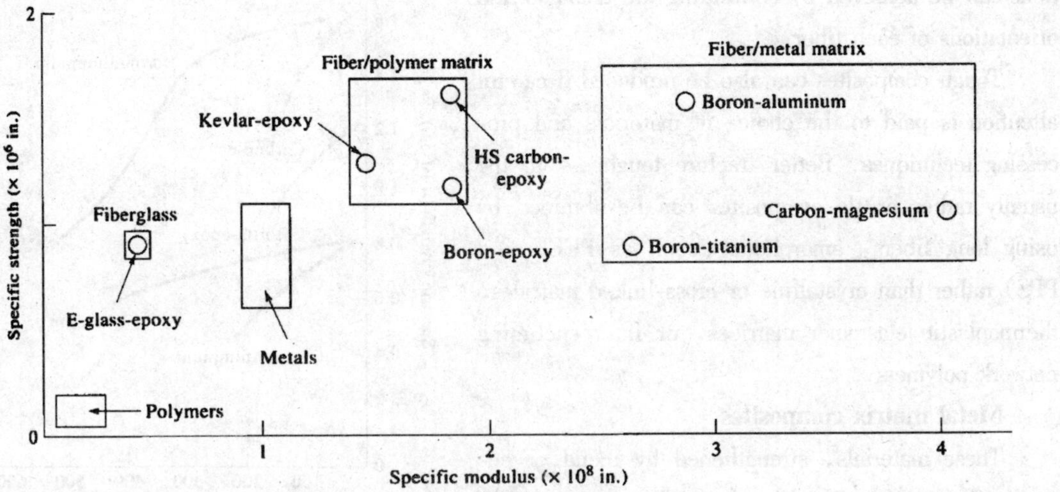

Figure 5.27 A comparison of the specific modulus and specific strength of several composite materials with those of metals and polymers

ly are polymer matrix composites reinforced with high-strength polymer, metal, or ceramic fibers. Carbon fibers are used extensively where particularly good stiffness is required; aramid — and, to an even greater extent, polyethylene — fibers are better suited to high-strength applications in which toughness and damage resistance are more import. Unfortunately, the polymer fibers lose their strength at relatively low temperatures, as do all of the polymer matrices (Figure 5.28).

Table 5.2 Examples of fiber-reinforced materials and applications

Materials	Applications
Borsic-aluminum	Fan blades in engines, other aircraft and aerospace applications
Kevlar-epoxy and Kevlar-polyester	Aircraft, aerospace applications (including space shuttle), boat hulls, sporting goods (including tennis rackets, golf club shafts, fishing rods), flak jackets
Graphite-polymer	Aerospace and automotive applications, sporting goods
Glass-polymer	Lightweight automotive applications, water and marine applications, corrosion-resistant applications, sporting goods equipment, aircraft and aerospace components

The advanced composites are also frequently used for sporting goods. Tennis rackets, golf clubs, skis, ski poles, and fishing poles often contain carbon or aramid fibers because the higher stiffness provides better performance. In the case of golf clubs, carbon fibers allow less weight in the shaft and therefore more weight in the head. Fabric reinforced with polyethylene fibers are used for light-weight sails for racing yachts.

A unique application for aramid fiber composites is armor. Tough Kevlar composites provide better ballistic protection than do other materials, making them suitable for lightweight, flexible bulletproof clothing.

Hybrid composites are composed of two or more types of fibers. For instance, Kevlar fibers may be mixed with carbon fibers to improve the toughness of a stiff composite, or Kevlar may be mixed with glass fibers to improve stiffness. Particularly good tailoring of the composite to meet specific applications can be achieved by controlling the amounts and orientations of each fiber.

Tough composites can also be produced if careful attention is paid to the choice of materials and processing techniques. Better fracture toughness in the usually rather brittle composites can be obtained by using long fibers, amorphous (such as PEEK and PPS) rather than crystalline or cross-linked matrices, thermoplastic elastomer matrices, or interpenetrating network polymers.

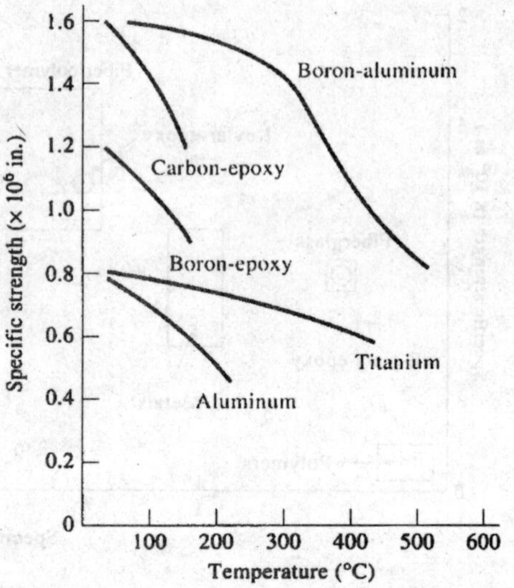

Figure 5.28 The specific strength versus temperature for several composites and metals

Metal matrix composites

These materials, strengthened by metal or ceramic fibers, provide high-temperature resistance. Aluminum reinforced with borsic fibers has been used extensively in aerospace applications, including struts for the space shuttle. Copper-base alloys have been reinforced with SiC fibers, producing high-strength propellers for ships.

Aluminum is commonly used in metal matrix composites. Al_2O_3 fibers reinforce the pistons for some diesel engines; SiC fibers and whiskers are used in aerospace applications, including stiffeners and missile fins; and carbon fibers provide reinforcement for the aluminum antenna mast of the Hubble telescope. Polymer fibers, because of their low melting or degradation temperatures, are not normally used in a metallic matrix. Polymets, however, are produced by hot-extruding aluminum powder and high-melting-temperature liquid-crystalline polymers. A reduction of 1,000 to 1 during the extrusion process elongates the polymer into aligned filaments and bonds the aluminum powder particles into a solid matrix.

Metal matrix composites may find important applications in components for rocket or aircraft engines. Superalloys reinforced with metal fibers (such as tungsten) or ceramic fibers (such as SiC or B_4N) maintain their strength at higher temperatures, permitting jet engines to operate more efficiently. Similarly, titanium and titanium aluminides reinforced with SiC fibers are considered for turbine blades and disks.

A unique application for metal matrix composites is in the superconducting wire required for fusion reactors. The intermetallic compound Nb_3Sn has good superconducting properties but is very brittle. To produce Nb_3Sn wire, pure niobium wire is surrounded by copper as the two metals are formed into a wire composite (Figure 5.29). The niobium-copper composite wire is then coated with tin. The tin diffuses through the copper and react with the niobium to produce the intermetallic compound. Niobium-titanium systems are also used.

Ceramic matrix composites

Composites containing ceramic fibers in a ceramic matrix are also finding applications. Two important uses will be discussed to illustrate the unique properties that can be obtained with these materials.

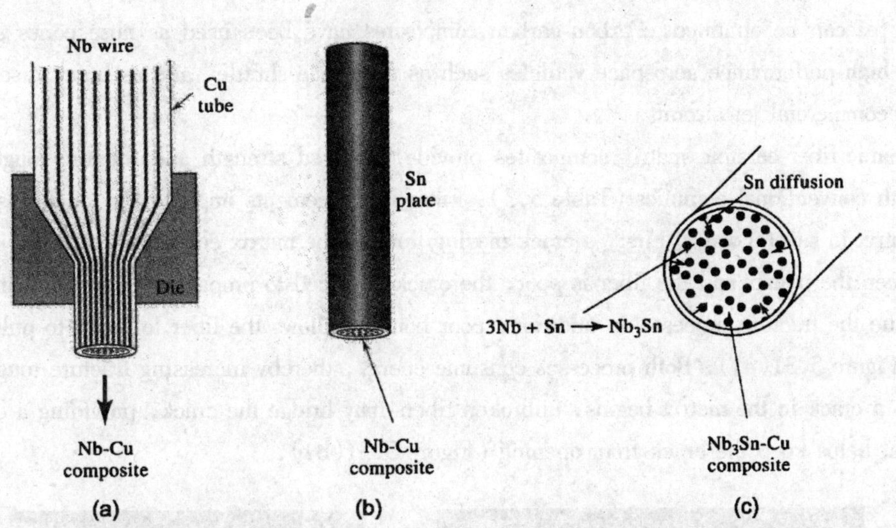

Figure 5.29 The manufacture of composite superconductor wires: (a) Niobium wire is surrounded with copper during forming. (b) Tin is planted onto Nb-Cu composite wire. (c) Tin diffuses to niobium to produce the Nb_3Sn-Cu composite

Carbon-carbon composites are used for extraordinary temperature resistance in aerospace applications. Carbon-carbon composites can operate at temperatures of up to 3,000 ℃ and, in fact, are stronger at high temperatures than at low temperatures (Figure 5.30). Carbon-carbon composites are made by forming a polyacrylonitrile or carbon fiber fabric into a mold, then impregnating the fabric with an organic resin, such as a phenolic. The part is pyrolyzed to convert the phenolic resin to carbon. The composite, which is still soft and porous, is impregnated and pyrolyzed several more times, continually increasing the density, strength, and stiffness. Finally the part is coated with silicon carbide to protect the carbon-carbon composite from oxidation. Strengths of 300,000 psi and stiffnesses of

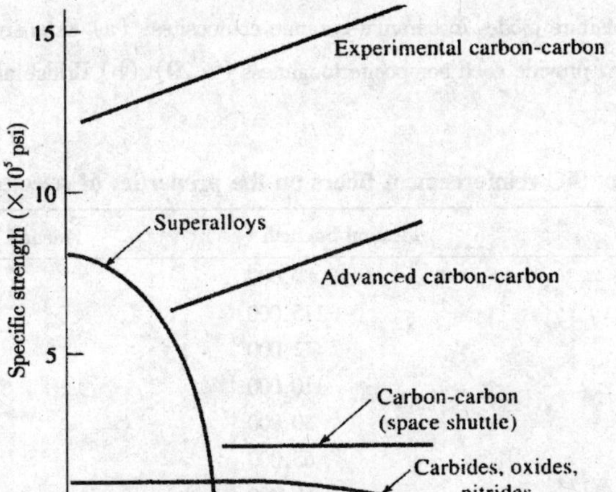

Figure 5.30 A comparison of the specific strength of various carbon-carbon composites with that of other high-temperature materials relative to temperature

50×10^6 psi can be obtained. Carbon-carbon composites have been used as nose cones and leading edges of high-performance aerospace vehicles such as the space shuttle, and as break discs on racing cars and commercial jet aircraft.

Ceramic fiber-ceramic matrix composites provide improved strength and fracture toughness compared with conventional ceramics (Table 5.3). Fiber reinforcements improve the toughness of the ceramic matrix in several ways. First, a crack moving through the matrix encounters a fiber; if the bonding between the matrix and the fiber is poor, the crack is forced to propagate around the fiber in order to continue the fracture process. In addition, poor bonding allows the fiber to begin to pull out of the matrix (Figure 5.31(a)). Both processes consume energy, thereby increasing fracture toughness. Finally, as a crack in the matrix begins, unbroken fibers may bridge the crack, providing a compressive stress that helps keep the crack from opening (Figure 5.31(b)).

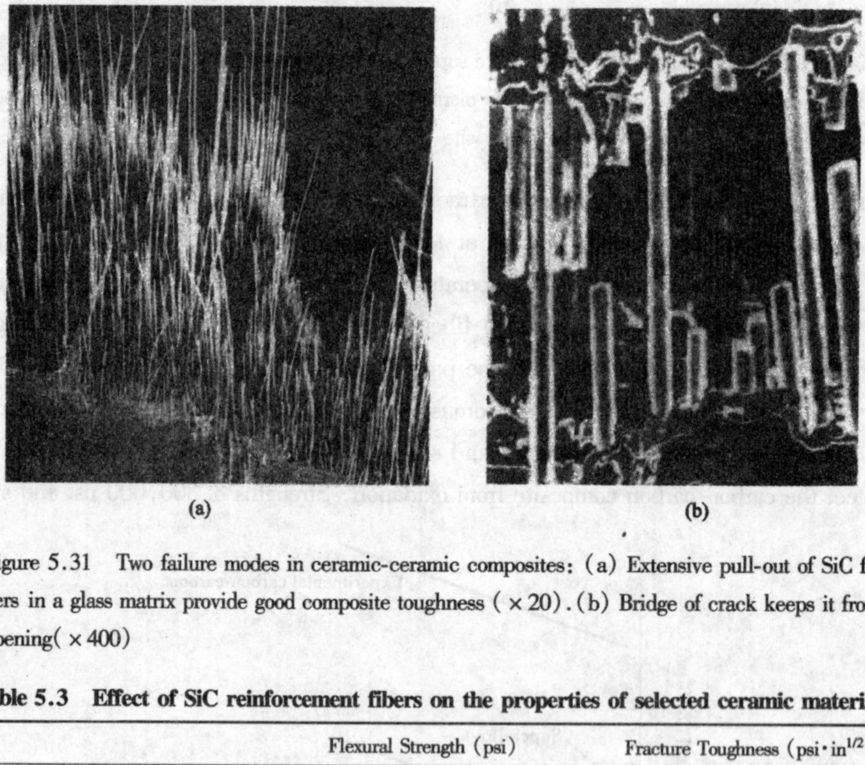

(a) (b)

Figure 5.31 Two failure modes in ceramic-ceramic composites: (a) Extensive pull-out of SiC fibers in a glass matrix provide good composite toughness ($\times 20$). (b) Bridge of crack keeps it from opening($\times 400$)

Table 5.3 Effect of SiC reinforcement fibers on the properties of selected ceramic materials

Materials	Flexural Strength (psi)	Fracture Toughness ($psi \cdot in^{1/2}$)
Al_2O_3	80,000	5,000
Al_2O_3/SiC	115,000	8,000
SiC	72,000	4,000
SiC/SiC	110,000	23,000
ZrO_2	30,000	5,000
ZrO_2/SiC	65,000	20,200
Si_3N_4	68,000	4,000
Si_3N_4/SiC	115,000	51,000
Glass	9,000	1,000
Glass/SiC	120,000	17,000
Glass ceramic	30,000	2,000
Glass ceramic/SiC	120,000	16,000

Unlike polymer and metal matrix composites, poor bonding — rather than good bonding — is required. Consequently, control of the interface structure is crucial. In a glass ceramic (based on $Al_2O_3 \cdot SiO_2 \cdot Li_2O$) reinforced with SiC fibers, an interface layer containing carbon and NbC is produced that makes debonding of the fiber from the matrix easy. If, however, the composite is heated to a high temperature, the interface is oxidized; the oxide occupies a large volume, exerts a clamping force on the fiber, and prevents easy pull-out. Fracture toughness is then decreased.

5.8 Laminar composite materials

Laminar composites include very thin coatings, thicker protective surfaces, claddings, bimetallics, laminates, and a host of other applications. In addition, the fiber-reinforced composites produced from tapes or fabrics can be considered as partly laminar. Many laminar composites are designed to improve corrosion resistance while retaining low cost, high strength, or light eight. Other important characteristics include superior wear or abrasion resistance, improved appearance, and unusual thermal expansion characteristics.

Rule of mixtures

Some properties of the laminar composite materials parallel to the lamellae are estimated from the rule of mixtures. Density, electrical and thermal conductivity, and modulus of elasticity can be calculated with little error:

$$\begin{aligned} &\text{Density} = \rho_c = \sum f_i \rho_i \\ &\text{Electrical conductivity} = \sigma_c = \sum f_i \sigma_i \\ &\text{Thermal conductivity} = K_c = \sum f_i K_i \\ &\text{Modulus of elasticity} = E_c = \sum f_i E_i \end{aligned} \tag{5.13}$$

The laminar composites are very anisotropic. The properties perpendicular to the lamellae are:

$$\begin{aligned} &\text{Electrical conductivity} = \frac{1}{\sigma_c} = \sum \frac{f_i}{\sigma_i} \\ &\text{Thermal conductivity} = \frac{1}{K_c} = \sum \frac{f_i}{K_i} \\ &\text{Modulus of elasticity} = \frac{1}{E_c} = \sum \frac{f_i}{E_i} \end{aligned} \tag{5.14}$$

However, many of the really important properties, such as corrosion and wear resistance, depend primarily on only one of the components of the composite, so the rule of mixtures is not applicable.

Producing laminar composites

Several methods are used to produce laminar composites, including a variety of deformation and joining techniques (Figure 5.32).

Individual plies are often joined by adhesive bonding, as is the case in producing plywood. Polymer matrix composites built up from several layers of fabric or tape prepregs are also joined by adhesive bonding; a film of unpolymerized polymer is placed between each layer of prepreg. When the layers

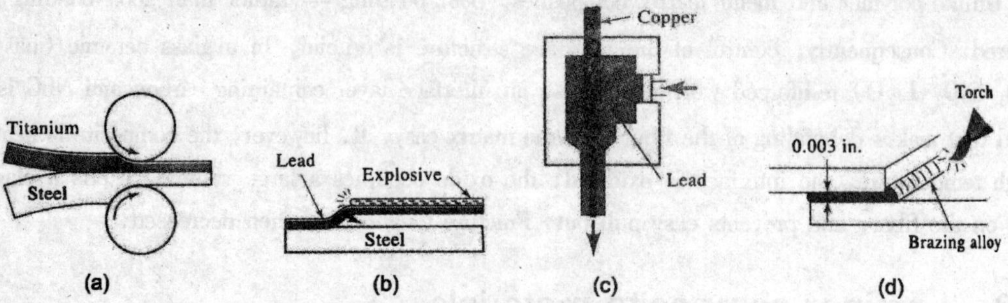

Figure 5.32 Techniques for producing laminar composites: (a) roll bonding, (b) explosive bonding, (c) coextrusion, and (d) brazing

are pressed at an elevated temperature, polymerization is completed and the prepregged fibers are joined to produce composites that may be dozens of layers thick.

Most of the metallic laminar composites, such as claddings and bimetallics, are produced by deformation bonding, such as hot or cold roll bonding. The pressure exerted by the rolls breaks up the oxide film at the surface, brings the surfaces into atom-atom contact, and permits the two surfaces to be joined. Explosive bonding can also be used. An explosive charge provides the pressure required to join metals. This process is particularly well suited for joining very large plates that will not fit into a rolling mill.

Very simple laminar composites, such as coaxial cable, are produced by coextruding two materials through a die in such a way that the soft material surrounds the harder material. Metal conductor wire can be coated with an insulting thermoplastic polymer in this manner.

Brazing can join composite plates. The metallic sheets are separated by a very small clearance — preferably, about 0.003 in. — and heated above the melting temperature of the brazing alloy. The molten brazing alloy is drawn into the thin joint by capillary action.

5.9 Examples and applications of laminar composites

The number of laminar composites is so varied and their applications and intentions are so numerous that we cannot make generalizations concerning their behavior. Instead we will examine the characteristics of a few commonly used examples.

Laminates

Laminates are layers of materials joined by an organic adhesive. In safety glass, a plastic adhesive, such as polyvinyl butyral, joins two pieces of glass; the adhesive prevents fragments of glass from flying about when the glass is broken. Laminates are used for insulation in motors, for gears, for printed circuit boards, and for decorative items such as Formica countertops and furniture.

Microlaminates include composites composed of alternating layers of aluminum sheet and fiber-reinforced polymer. Arall (aramid aluminum laminate) and Glare (glass aluminum laminate) have been developed as possible skin materials for aircraft. In Arall, an aramid fiber such as Kevlar is prepared as a fabric or unidirectional tape, impregnated with an adhesive, and laminated between layers of aluminum alloy (Figure 5.33). The composite laminate has an unusual combination of strength, stiffness, corrosion resistance, and light weight. Fatigue resistance is improved, since the interface between the layers may block cracks. Compared with polymer matrix composites, the microlaminates have good resistance to lighting strike damage (which is important in aerospace applications), are formable and machinable, and are easily repaired.

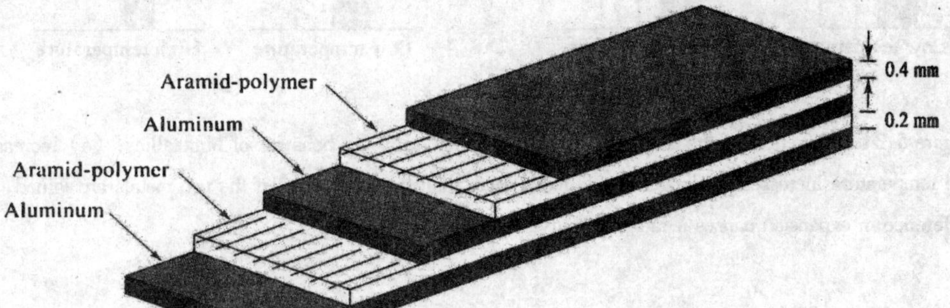

Figure 5.33 Schematic diagram of an aramid aluminum laminate, Arall, which has potential for aerospace applications

Clad metals

Clad materials are metal-metal composites. A common example of cladding is United States silver coinage. A Cu-80% Ni alloy is bonded to both sides of a Cu-20% Ni alloy. The ratio of thickness is about 1/6:2/3:1/6. The high-nickel alloy is a silver color, while the predominantly copper core provides low cost.

Clad materials provide a combination of good corrosion resistance with high strength. Alclad is a clad composite in which commercially pure aluminum is bonded to higher-strength alloy from corrosion. The thickness of the pure aluminum layer is about 1% to 15% of the total thickness. Alclad is used in aircraft construction, heat exchangers, building construction, and storage tanks, where combinations of corrosion resistance, strength, and light weight are desired.

Bimetallics

Temperature indicators and controllers take advantage of the different coefficients of thermal expansion of the two metals in laminar composite. If two pieces of metal are heated, the metal with the higher coefficient of thermal expansion becomes longer (Figure 5.34). If the two pieces of metal are rigidly bonded together, the difference in their coefficients causes the strip to bend and produce a curved surface. The amount of movement depends on the temperature; by measuring the curvature or deflection of the strip, we can determine the temperature. Likewise, if the free end of the strip activates a relay, the strip can turn on or off a furnace or air conditioner to regulate temperature.

Metals selected for bimetallics must have ① very different coefficients of thermal expansion, ② expansion characteristics that are reversible and repeatable, and ③ a high modulus of elasticity, so that the bimetallic device can do work. Often the low-expansion strip is made from Invar, an iron-nickel alloy, whereas the high-expansion strip may be brass, Monel, or pure nickel.

Bimetallics can act as circuit breakers as well as themostats; if a current passing through the strip becomes too high, heating causes the bimetallic to deflect and break the circuit.

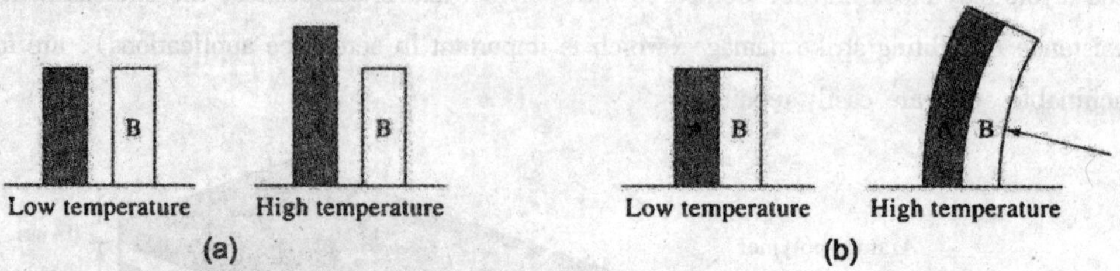

Figure 5.34 The effect of the thermal expansion coefficient on the behavior of bimetallics: (a) increasing the temperature increases the length of one metal more than the other, (b) if the two metals are joined, the difference in expansion causes a radius of curvature to be produced

5.10 Sandwich structures

Sandwich materials have thin layers of a facing material joined to a lightweight filler material, such as a polymer foam. Neither the filler nor the facing material is strong or rigid, but the composite possesses both properties. A familiar example is corrugated cardboard. A corrugated core of paper is bonded on either side to flat, thick paper. Neither the corrugated core nor the facing paper is rigid, but the combination is.

Another important example is the honeycomb structure used in aircraft applications. A honeycomb is produced by gluing thin aluminum strip at selected locations. The honeycomb material is then expanded to produce a very low density cellular panel that, by itself, is unstable (Figure 5.35). When an aluminum facing sheet is adhesively bonded to either side of the honeycomb, however, a very stiff, rigid, strong, and exceptionally lightweight sandwich with a density as low as 0.04 g/cm^3 is obtained.

The honeycomb cells can have a variety of shapes, including hexagonal, square, rectangular,

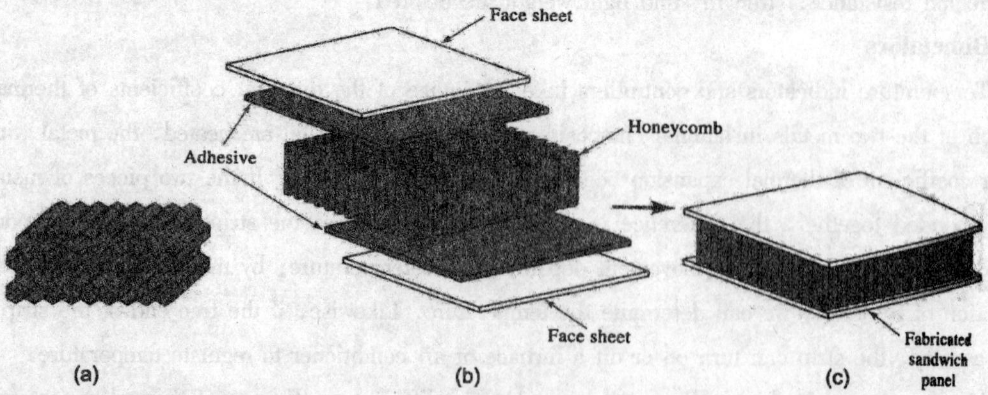

Figure 5.35 A hexagonal cell honeycomb core (a) can be joined to two face sheets by means of adhesive sheets (b), producing an exceptionally lightweight yet stiff, strong honeycomb sandwich structure (c)

and sinusoidal, and they can be made from aluminum, fiberglass, paper, aramid polymers, and other materials. The honeycomb cells can be filled with foam or fiberglass to provide excellent sound and vibration absorption. Figure 5.36 describes one method by which the honeycomb can be fabricated.

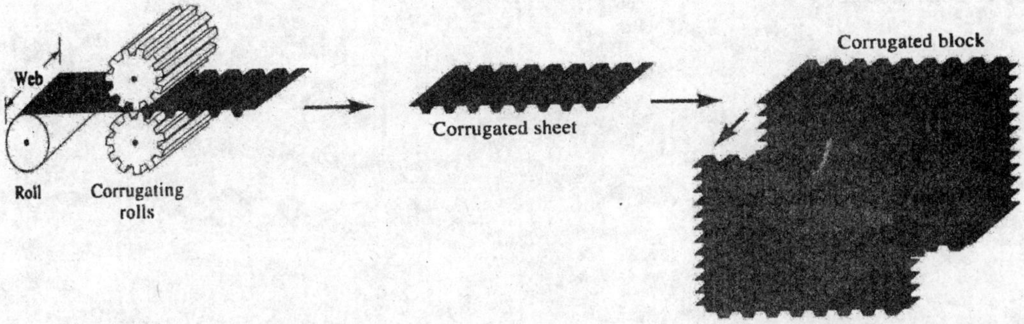

Figure 5.36 In the corrugation method for producing a honeycomb core, a roll of the material (such as aluminum) is corrugated between two rolls. The corrugated sheets are joined together with adhesive and then cut to the desired thickness

Keywords

aramid fibers
aspect ratio
carbonizing
cemented carbides
dispersoids
filament winding
honeycomb
pultrusion
precursor
rule of mixtures
specific strength
sandwich
specific modulus
whiskers

Problems

1. Calculate the density of a cemented carbide, or cermet, based on a titanium matrix if the composite contains 50 %(wt) WC, 22 %(wt) TaC, and 14 %(wt) TiC. (9.408 g/cm^3)

2. How much clay must be added to 10 kg of polyethylene to produce a low-cost composite having a modulus of elasticity greater than 120,000 psi and a tensile strength greater than 2,000 psi? The density of the clay is 2.4 g/cm^3 and that of polyethylene is 0.92 g/cm^3. (11.18 to 22.2 kg)

3. Explain why bonding between carbon fibers and epoxy matrix should be excellent, whereas bonding between silicon nitride fibers and a silicon carbide matrix should be poor.

Selected from

[1] D. R. Askeland. The Science and Engineering of Materials. 3rd Edition. Boston: PWS Publishing Company, 1996.

[2] J. F. Shackelford. Introduction to Materials Science for Engineers. 2nd Edition. New York: Macmillan Publishing Company, 1988.

[3] W. D. Callister. Fundamentals of Materials Science and Engineering. 5th Edition. John Wiley & Sons, 2001.